Geology of the Glenwood Springs Quadrangle of North Western Colorado

by US Dept. of Interior

with an introduction by Kerby Jackson

This work contains material that was originally published in 1963.

This publication is within the Public Domain.

This edition is reprinted for educational purposes and in accordance with all applicable Federal Laws.

Introduction Copyright 2016 by Kerby Jackson

Introduction

It has been years since the important publication "Geology of the Glenwood Springs Quadrangle and Vicinity of North Western Colorado" was published. First released in 1963, this work has been unavailable to the mining community since those days, with the exception of expensive original collector's copies and poorly produced digital editions.

It has often been said that "*gold is where you find it*", but even beginning prospectors understand that their chances for finding something of value in the earth or in the streams of the Golden West are dramatically increased by going back to those places where gold and other minerals were once mined by our forerunners. Despite this, much of the contemporary information on local mining history that is currently available is mostly a result of mere local folklore and persistent rumors of major strikes, the details and facts of which, have long been distorted. Long gone are the old timers and with them, the days of first hand knowledge of the mines of the area and how they operated. Also long gone are most of their notes, their assay reports, their mine maps and personal scrapbooks, along with most of the surveys and reports that were performed for them by private and government geologists. Even published books such as this one are often retired to the local landfill or backyard burn pile by the descendents of those old timers and disappear at an alarming rate. Despite the fact that we live in the so-called "Information Age" where information is supposedly only the push of a button on a keyboard away, true insight into mining properties remains illusive and hard to come by, even to those of us who seek out this sort of information as if our lives depend upon it. Without this type of information readily available to the average independent miner, there is little hope that our metal mining industry will ever recover.

This important volume and others like it, are being presented in their entirety again, in the hope that the average prospector will no longer stumble through the overgrown hills and the tailing strewn creeks without being well informed enough to have a chance to succeed at his ventures.

Please note that at times it is necessary to rearrange illustration plates in these texts. Any illustrations not found in their original sequence may be found following the index. Regrettably, at this time, we cannot reproduce the large fold out maps that sometimes appeared in the original edition.

Kerby Jackson
June 2016

www.goldminingbooks.com

CONTRIBUTIONS TO ECONOMIC GEOLOGY

GEOLOGY OF THE GLENWOOD SPRINGS QUADRANGLE AND VICINITY, NORTHWESTERN COLORADO

By N. Wood Bass and Stuart A. Northrop

ABSTRACT

The Glenwood Springs quadrangle and vicinity covers an area of about 900 square miles in parts of Garfield, Eagle, Routt, and Rio Blanco Counties in northwestern Colorado. The map area includes much of the White River uplift, where rocks ranging in age from Precambrian to Quaternary are exposed. These exposed sedimentary rocks and lava flows have a total thickness of about 24,800 feet. The general structure of the area is shown on the map by structure contours drawn on top of the Leadville Limestone at intervals of 500 feet. The uplift is characterized by many reverse and tension faults, many of which trend slightly north of west at about right angles to the trend of reverse faults elsewhere in the State.

Several thick beds of bituminous coal, which occur in the Mesaverde Group of Late Cretaceous age, crop out in the southwestern part of the map area. Several formations which crop out in the map area contain beds that are prospectively valuable as reservoirs for oil and gas outside the area of this report. Other deposits of economic importance include four beds of gypsum, which occur in the Paradox Formation and range from 65 to 160 feet in thickness, and thick units of high-grade limestone, which crop out on the Colorado River near the Denver and Rio Grande Western Railroad. Volcanic cinders are quarried and used in the manufacture of cinder block and for road metal. Travertine and volcanic ash are potentially valuable.

INTRODUCTION

The Glenwood Springs quadrangle and small parts of the adjacent area are in northwestern Colorado and include parts of Garfield, Eagle, Routt, and Rio Blanco Counties. (See fig. 1.) The quadrangle extends from long. 107°00′ to 107°30′ W. and lat. 39°30′ to 40°00′ N. It includes a large part of the White River uplift, where rocks ranging in age from Precambrian to Quaternary are exposed. Figure 1 shows the main physiographic features of Colorado. The stippled areas represent the mountain ranges, which are large anticlines, elongate northwestward, separated by major synclines. The White River uplift seems to be a northwestward extension of the Sawatch Mountains, but slightly en echelon to them.

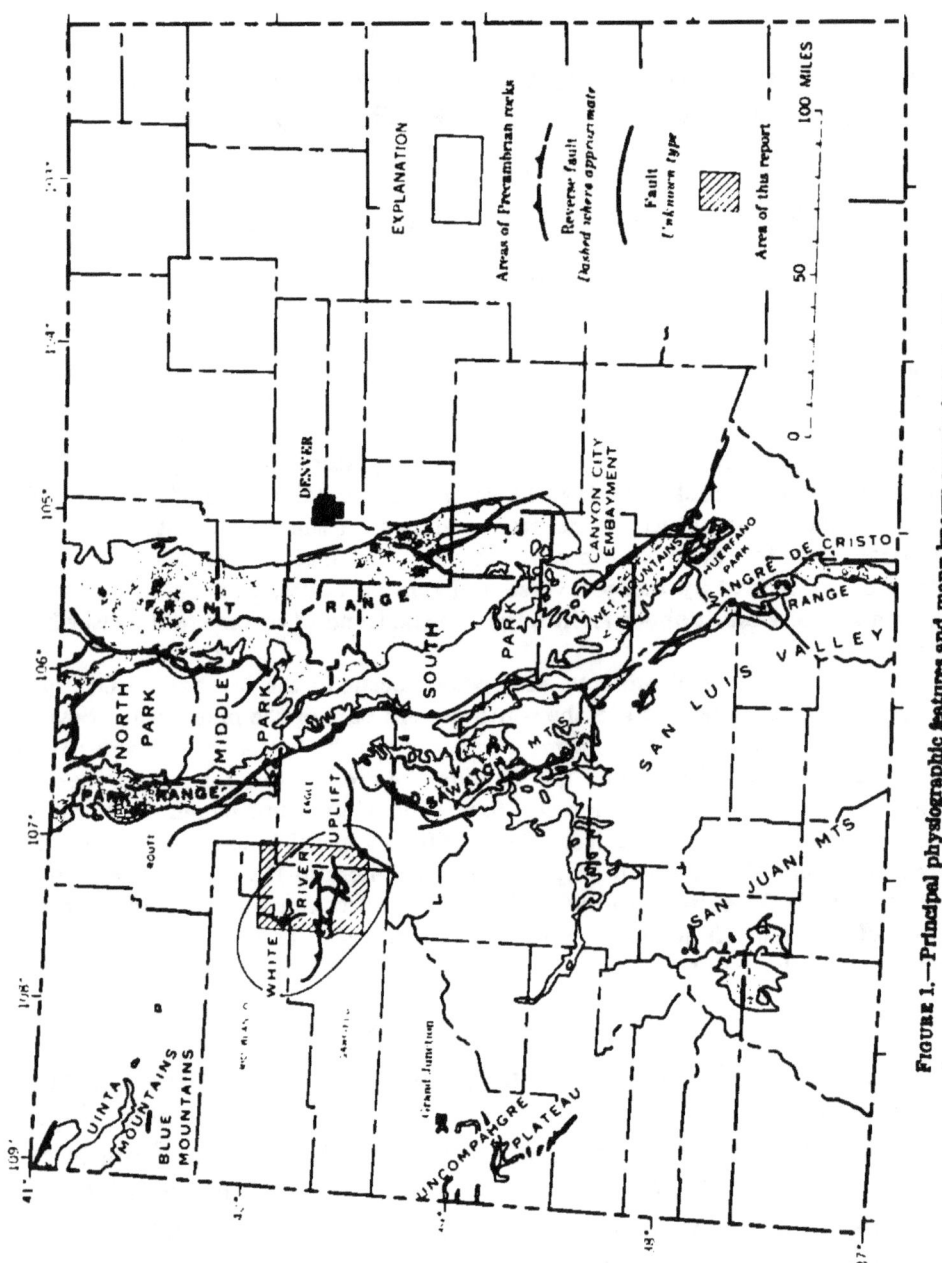

FIGURE 1.—Principal physiographic features and many known reverse faults in Colorado.

Altitudes within the quadrangle range from about 5,600 feet, where the Colorado River leaves the west edge of the quadrangle, to 12,246 feet, on Sheep Mountain. Much of the White River Plateau exceeds 10,000 feet in altitude; much of the Flat Tops area ranges from 10,500 to 11,500 feet. Several peaks in the northeast quarter of the quadrangle exceed 12,000 feet. In the north half of the quadrangle are many small lakes.

The principal streams in the map area are the Colorado and Eagle Rivers in the south part, the South Fork of the White River in the north part, and their tributaries. The Denver and Rio Grande Western Railroad follows the Colorado and Eagle Rivers, as does U.S. Highway 6 and 24.

Sedimentary rocks and lava flows in the map area have a total thickness of about 24,800 feet. (See pl. 2.) This total includes about 6,930 feet of rocks of Paleozoic age, 11,270 feet of rocks of Mesozoic age, and 6,600 feet of rocks of Cenozoic age.

Commercially valuable coal beds in the Mesaverde Group of Late Cretaceous age crop out in the southwest part of the quadrangle. Several of the formations exposed in the area are prospectively valuable as reservoirs for oil and gas in the Uinta Basin and elsewhere in northwestern Colorado. Hence, this investigation was conducted mainly to study the rocks in outcrop so that the data obtained can be applied to the surrounding regions, where these rocks are deeply buried.

FIELD AND OFFICE WORK

The area was mapped in the field on aerial photographs at a scale of about 1:24,000; data obtained were later transferred to a base map, whose scale is 1:31,680. Some trails shown on the map were sketched on photographs, and others were transferred from maps by the U.S. Forest Service (Watts and Dean, 1949).

The fieldwork was done during parts of seven field seasons, from 1947 to 1953. The senior author was in the field each season and participated in the measuring of stratigraphic sections, collection of fossils, and geologic mapping. The junior author was in the field during part of three seasons, during which time he measured many of the stratigraphic sections and made many of the fossil collections; the fossil determinations for the Pennsylvanian, the Permian, and some of the Devonian collections were made by him. The authors are grateful to the following geologists who aided in the fieldwork: James D. Vine, Frank G. Cooley, Harley F. Barnes, James K. Weaver, Raymond C. Robeck, Peter M. Thompson, and the late John C. Benson; and to Mrs. Flora K. Walker and Mrs. Marjorie H. Barnett, who prepared the planimetric map. Allison R. Palmer made fossil collections in the field and with others identified the fauna from the Cambrian and Ordovician rocks. Several paleontologists made determinations of fossil collections that were sent them by us; these include the late John B. Reeside, Jr., D. H. Dunkle, the late W. H. Hass, the late Josiah Bridge, G. A. Cooper, the late J. B. Knight, Helen Duncan, I. G. Sohn, W. A. Cobban, P. E. Cloud, Jr., J. Harlan Johnson, Lloyd G. Henbest, and Thomas G. Roberts. Ogden L. Tweto identified the boundary between the Devonian and Mississippian rocks in the field. G. M. Richmond made a reconnaissance examination of the glacial deposits and prepared an informal report.

Certain aspects of the geology of the area have been published previously by the authors (Bass and Northrop, 1950, 1953, and 1955; Bass, 1956, 1958).

METAMORPHIC AND IGNEOUS ROCKS

PRECAMBRIAN ROCKS

Precambrian rocks are exposed chiefly in the deep canyons of the Colorado River and its tributaries and of the South Fork of the White River. They consist of early quartz-biotite schists and greenstones that are intruded by granites and pegmatite dikes. MacQuown (1945, p. 881) believes the schist to have been originally sedimentary, and T. S. Lovering (personal communication to MacQuown) suggested that the schists probably are equivalent to the Idaho Springs Formation of the Front Range in Colorado. MacQuown (1945, p. 884) found that the strike of the schistosity is N. 70° W., which suggested to him the presence in Precambrian time of a northwestward structural trend in northwestern Colorado, extending at least from the White River uplift to the Uinta Mountains.

SEDIMENTARY ROCKS

UPPER CAMBRIAN

A sequence of quartzite, sandstone, dolomite, and limestone flat-pebble conglomerate interbedded with very thin beds of shale is exposed widely in the map area. The sequence is assigned tentatively a Late Cambrian age. The total thickness of the sequence is about 600 feet; the Sawatch Quartzite includes the lowermost 500 feet and the Dotsero Formation includes the uppermost 100 feet.

SAWATCH QUARTZITE

The Sawatch Quartzite consists chiefly of regularly bedded sandstone, quartzitic sandstone, and quartzite, in beds commonly ranging from 2 to 5 feet in thickness. In addition, the formation contains a few units of thin-bedded dolomite. All these are interbedded with beds of light greenish-gray shale that range from a fraction of an inch to several inches in thickness. The regular bedding of the formation is one of its chief characteristics.

A unit 75 feet or more in thickness, composed of dark-brown thin-bedded dolomite and sandy dolomite containing much glauconite, is present in the upper part of the formation. The position of the dolomite unit below the top of the formation ranges from 65 to 150 feet. Other units of dolomite, whose thickness ranges from 4 to 17 feet, are present in the Sawatch at some places. The beds of quartzite are most abundant in the part of the formation above the 75-foot dolomite; these quartzite beds are very light gray and give the cliff face a banded light-and-dark appearance. The only fossils found in the Sawatch were a very few chitinophosphatic brachiopods in this dolomite unit.

The Sawatch Quartzite forms sheer cliffs 400 to 500 feet high in Glenwood Canyon, and cliffs nearly as high in the canyons of Deep, Grizzly, and Canyon Creeks, and the South Fork of the White River. The 75-foot dolomite unit forms a notch or shoulder in the cliffs and supports a scant growth of pine trees and many shrubs. The contact of the formation with the underlying Precambrian rocks is sharp at the few places where it is exposed. The boundary between the Sawatch Quartzite and the overlying Dotsero Formation is defined by relatively thick beds of quartzite below, and shale and thin beds of dolomite above. Inasmuch as the Dotsero forms a slope that recedes from the edge of the cliff of the Sawatch, the contact is readily distinguishable in mapping. Most of the large springs on the White River Plateau issue from beds in the Sawatch Quartzite.

A stratigraphic section of the Sawatch Quartzite, measured in Glenwood Canyon, is described below. This section is not wholly typical of the formation in much of the map area in that here it contains a larger content of quartzite than at most other places and the purple color of many beds here is a local feature. A light-brown color characterizes the formation nearly everywhere else in the map area.

Stratigraphic section of the Sawatch Quartzite, measured by Harley F. Barnes in the cliffs on the north side of U.S. Highway 6, 3.9 miles by highway southwest of the Eagle-Garfield County line, north of the center of sec. 20, T. 5 S., R. 87 W.

Dotsero Formation (Upper Cambrian).
Sawatch Quartzite (Upper Cambrian):

	Ft	In
22. Quartzite, dolomitic, micaceous, very glauconitic, fine-grained, light-brown	5	0
21. Covered. Probably same as above and below	11	9
20. Quartzite, dolomitic, very glauconitic, micaceous, fine-grained, light-brown	2	0
19. Quartzite, fine-grained, purplish-white	10	0

Sawatch Quartzite (Upper Cambrian)—Continued

	Ft	In
18. Dolomite, sandy, fine-grained, buff-white-----------------	1	0
17. Quartzite, calcareous, fine-grained; beds of denser quartzite are whiter than other beds and impart to this unit a banded character which persists throughout the map area.	55	0
16. Dolomite, finely crystalline, buff-gray---------------------	4	0
15. Quartzite, fine-grained, white, weathers light brown; forms conspicuous light band on face of cliff-----------------	7	0
14. Dolomite, sandy, finely crystalline, greenish-buff----------	4	0
13. Quartzite, fine-grained; alternate white and light-brown beds give the unit a banded character------------------	25	0
12. Dolomite, sandy, finely crystalline, glauconitic, finely cross-bedded, greenish-buff-----------------------------	9	0
11. Quartzite, dolomitic, fine-grained, micaceous, glauconitic, thin- to thick-bedded, light-greenish-brown; includes beds of quartzitic dolomite and one 2-ft bed of massive quartzite--	16	0
10. Dolomite, finely crystalline; some beds are sandy, glauconitic, and micaceous, especially in upper part of unit where parts grade into dolomitic sandstone; thin- to thick-bedded; greenish-brown; characteristically forms a prominent notch in the cliff face; the bench of its base supports a scant growth of evergreen trees------------	69	0

Section offset 0.2 mile westward

	Ft	In
9. Quartzite, fine-grained, in beds 3 to 6 ft or more thick; buff to brown---	68	0
8. Dolomite, sandy, finely crystalline, glauconitic, in beds as much as 4 in. thick; buff; includes local lenses of quartzite; forms a notch in cliffs-------------------------	11	0
7. Quartzite, fine-grained, white, generally thick-bedded; forms sheer cliff, prominently jointed; weathering has developed chimneys 3 to 5 ft wide, 20 to 50 ft long, 70 ft high.	77	0
6. Sandstone, locally arkosic, fine- to medium-grained, crossbedded, chiefly rounded grains, calcareous cement in part; forms a bench which interrupts sheer cliff; weathers light tan, white, and purplish------------------	22	0
5. Quartzite, arkosic, glauconitic; grain size ranges from fine to very coarse; calcareous cement present locally, siliceous cement most common; crossbedded within beds ranging in thickness from 1 in. to 2 ft; light-brown, but weathered rock in cliff face is banded alternately purplish and white.	117	0
4. Shale, very micaceous, purple----------------------------		4
3. Quartzite, arkosic, medium-grained, crossbedded, purple to buff---	2	0
2. Quartzite conglomerate, purple; angular fragments are as much as 5×3×3 in-------------------------------------		6
1. Arkose, coarse- to medium-grained, siliceous, very ferruginous, purplish-red; angular quartz and feldspar fragments as large as 1 in. in diameter-------------------		5
Total thickness of Sawatch Quartzite---------------------	517	0

Unconformity.

Precambrian:

    Granite, foliated, cut by pegmatites. Top 10 ft weathered.

UPPER CAMBRIAN AND LOWER ORDOVICIAN

Overlying the Sawatch Quartzite is a sequence of rocks, 185 to 250 feet thick, consisting of interbedded flat-pebble limestone and dolomite conglomerate, dolomite, and minor amounts of shale of Late Cambrian and Early Ordovician age. The sequence is divisible into three main lithologic units: (a) a lowermost unit comprising about one-fourth of the total thickness, consisting of thin-bedded tan dolomite and a few beds of flat-pebble dolomite conglomerate interbedded with greenish-tan dolomitic shale; (b) a middle unit comprising about one-half of the total, consisting mainly of thin beds of gray to grayish-tan flat-pebble limestone conglomerate that weathers to a reddish cast, interbedded with greenish-gray limy shale; and (c) an uppermost unit, comprising about one-fourth of the total, consisting of thin beds of regularly bedded tan dolomite that commonly forms a cliff. These lithologic units, however, do not coincide with the named stratigraphic units. The sequence is divided into two formations of about equal thickness—the Dotsero Formation of Late Cambrian age and the Manitou Formation of Early Ordovician age. The Dotsero Formation and overlying Manitou Formation are shown on the geologic map (pl. 1) as a single unit.

Stratigraphic section of the Dotsero and Manitou Formations, measured in Glenwood Canyon, in the SE¼ sec. 16, T. 5 S., R. 87 W., 0.4 mile west of the White River National Forest boundary

Chaffee Formation (Upper Devonian):
    Parting Member (basal bed only):

	Feet
53. Dolomite, sandy, medium- to coarse-grained, buff-gray; rounded to subangular quartz grains; two to three beds__	2. 3

Manitou Formation (Lower Ordovician):
    Tie Gulch Dolomite Member (type section):

52. Dolomite, thick-bedded, light-gray, weathering buff; dense; forms top of cliffs and a bench_____	6. 0
51. Dolomite, very fine to fine-grained, gray, grading laterally from thin bedded to massive_____	41. 7
Thickness of Tie Gulch Dolomite Member_____	47. 7

Dead Horse Conglomerate Member:

50. Shale, gray_____	. 1
49. Limestone, massive, gray, with thin shale beds at base; grades laterally into alternating thin-bedded limestone and shale_____	3. 2
48. Shale, gray, with thin limestone lenses_____	. 4
47. Limestone, thin-bedded, gray, with several shale partings__	6. 3

Manitou Formation (Lower Ordovician)—Continued
Dead Horse Conglomerate Member—Continued   *Feet*

46. Shale, gray, 0.1–0.2 ft thick_____ .2
45. Limestone, massive bed, fine-grained, gray_____ 2.0
44. Shale, gray, with thin limestone laminae_____ .6
43. Limestone, thin-bedded, fine- to medium-grained, gray;
      breccia at top_____ 3.0
42. Shale and thin-bedded limestone, interbedded_____ .5
41. Limestone conglomerate, thin- to medium-bedded, gray;
      weathers shaly in places and massive elsewhere_____ 24.5
   (*Fucoid markings or burrows are common on lower surfaces
   of beds from here down to unit 19.*)
40. Limestone, thin-bedded, medium crystalline, gray, with
      partings weathering yellow; contains rusty vugs_____ 7.0
39. Limestone conglomerate, massive to thick-bedded, finely
      crystalline, gray; contains vugs filled with pink calcite 5
      ft below top; micaceous 14 ft below top_____ 36.0
38. Shale, green, with thin limestone lenses_____ .7
37. Limestone, single massive bed, glauconitic, medium- to
      coarse-grained, light-gray and brown, with olive-green
      silty shale partings; upper 2–3 ft is dolomitic and lower
      6 ft is nodular and possibly conglomeratic_____ 12.8
36. Limestone conglomerate, fine-grained matrix, brown and
      gray; forms massive ledge_____ 4.5
35. Limestone, thin-bedded, weathering reddish, with shale part-
      ings and shaly weathering; in upper half, limestone beds
      about one-fourth in. thick interbedded with thinner shale
      laminae; two limestone lenses in lower half_____ 4.0
34. Limestone conglomerate, gray and brown; most pebbles are
      gray limestone, some red_____ .9
33. Shale, micaceous, maroon, interbedded with green shale;
      some fine- to medium-grained maroon-stained limestone
      interbedded as lenses_____ 1.2
                                                               _____
        Thickness of Dead Horse Conglomerate Member_____ 107.9
                                                               ======

        Total thickness of Manitou Formation_____ 155.6
Dotsero Formation (Upper Cambrian):
Clinetop Algal Limestone Member:
32. Limestone, algal or stromatolitic, single massive bed, brown
      and gray, weathering yellow and lavender to purple; base
      is wavy; in places the upper 1–2 ft is limestone conglomer-
      ate, containing abundant glauconite and weathering with
      a reddish cast; thickness of member ranges from 4.6 to
      5.0 ft_____ 5.0
                                                               ======
Glenwood Canyon Member (type section):
31. Shale, light greenish gray, and interbedded gray limestone;
      limestone weathers shaly; thickness ranges from 0.7 to
      1.0 ft_____ 1.0
30. Limestone conglomerate, in beds about 2 ft thick; both ma-
      trix and pebbles gray; several greenish partings; top bed
      stained red_____ 8.0

Dotsero Formation (Upper Cambrian)—Continued
Glenwood Canyon Member (type section)—Continued   *Feet*

29. Limestone conglomerate, massive; matrix and pebbles gray   5. 0

28. Shale and limestone, interbedded; limestone is gray, fine grained; shale is green gray_____   . 7

27. Limestone conglomerate, thin- to medium-bedded, medium- to fine-grained, mostly fine-grained, gray, weathering gray and buff_____   5. 8

26. Limestone, fine-grained, green-gray, weathering shaly_____   . 8

25. Limestone conglomerate, edgewise type, with flat pebbles as much as 6 in. across; fine-grained, gray_____   1. 2

24. Shale, hard, brittle, calcareous, green-gray_____   . 7

23. Limestone conglomerate and limestone, fine- to medium-grained; gray, mottled gray and yellowish buff, pebbles stained rusty yellow_____   2. 9

22. Dolomite, weathering shaly_____   . 4

21. Limestone conglomerate, thick-bedded, weathering gray and buff; matrix is medium grained; limestone pebbles, mostly flat, as much as 6 in. in diameter; upper 7–8 ft has gray limestone pebbles in gray limestone matrix; lower 3 ft has gray limestone pebbles in tan dolomitic matrix_____   10. 9

20. Shale, glauconitic, green-gray, and interbedded light-gray dolomite_____   1. 5

19. Conglomerate, medium-grained, glauconitic, weathering gray and rusty; dolomite pebbles mostly gray but upper surface has red pebbles; lower surfaces of beds greenish gray, marked by fucoids or burrows_____   2. 6

(Fucoid markings or burrows are common on lower surfaces of beds from here up to unit 41.)

18. Shale and dolomite, interbedded, with wavy top and base; dolomite is fine grained, tan; shale is fissile, green gray; unit contains one 8-in. lens of typical "red-cast" conglomerate and several rounded bioherm-like masses of dolomite as much as 9 in. high and 2–3 ft in diameter_____   2. 9

17. Dolomite, massive, fine-grained, gray; weathering rusty tan_   2. 5

16. Dolomite, weathering shaly_____   . 3

15. Dolomite and limestone conglomerate, "red-cast"; matrix is tan fine-grained calcitic dolomite; pebbles are dark red, mostly limestone, rounded, less than 1 in. across_____   . 85

14. Dolomite, in very thin beds with some green shale_____   . 85

13. Dolomite, thin-bedded to laminated, fine-grained; weathering buff to tan with white streaks, several greenish glauconitic beds_____   3. 7

12. Dolomite, thick-bedded to massive, medium-grained, gray and tan; basal 1-ft bed is variegated green and purple, weathering shaly_____   5. 8

11. Dolomite, medium-bedded, calcitic, fine- to medium-grained, green and tan streaked_____   5. 5

10. Dolomite, shaly; as thick as 1 ft_____   1. 0

9. Dolomite, lenticular, thin-bedded, wavy-bedded, cross-bedded, slightly calcitic, glauconitic, fine-grained, buff with green streaks_____   4. 8

Dotsero Formation (Upper Cambrian)—Continued
  Glenwood Canyon Member (type section)—Continued   *Feet*

    8. Dolomite, similar to unit 9_____ 6. 0

    7. Dolomite, lenticular, thin-bedded, wavy-bedded throughout, slightly calcitic, fine-grained, buff with green glauconitic streaks; upper foot and lower foot weathering shaly_____ 5. 3

    6. Dolomite, very glauconitic, coarse-grained, gray-yellow, weathering brighter yellow; many black grains_____ 2. 3

    5. Dolomite, very fine grained, weathering shaly; abundant black grains_____ . 3

    4. Dolomite, massive, grading laterally into thin- to medium-bedded phases, silty to very sandy; very glauconitic, very micaceous; fine-grained; gray, weathering brown gray, stained dark red on joint surfaces; abundant tiny grains of dark minerals; some thin shaly streaks_____ 7. 1

            Thickness of Glenwood Canyon Member_____ 90. 7

            Total thickness of Dotsero Formation_____ 95. 7

Sawatch Quartzite (Upper Cambrian) (only top 3 beds shown):

    3. Sandstone, crossbedded, calcitic, medium-grained, buff_____ 1. 5

    2. Sandstone, lenticular, thin- to medium-bedded, very dolomitic, medium-grained, with streaks rich in dark minerals, gray-tan; basal 1-ft bed makes setback, weathering shaly, but no shale is present_____ 3. 1

    1. Quartzite, alternating thin and thick lenses, slightly calcitic, fine- to medium-grained, light-tan, weathering yellowish to light brown_____ 3. 0

    (About 42 ft more of the Sawatch Quartzite is exposed at this locality.)

DOTSERO FORMATION

The Dotsero Formation includes the sequence of beds, 96 to 106 feet thick, directly above the Sawatch Quartzite; it consists of tannish-gray thin-bedded dolomite, flat-pebble limestone conglomerate, and a few beds of flat-pebble dolomite conglomerate, all interbedded with very thin beds of greenish-gray dolomitic shale. The formation is divisible into two members—the Glenwood Canyon Member, including all but the top few feet of the formation, and the Clinetop Algal Limestone Member, which embraces the top few feet of beds.

The basal contact of the Dotsero Formation is fairly sharply defined by the ledge- and cliff-forming quartzite beds of the underlying Sawatch Quartzite. Locally, however, the basal beds of the Dotsero Formation are sandy. The precise position of the upper contact of the Dotsero was determined in Glenwood Canyon where a Late Cambrian fauna was collected from a key bed in the Clinetop Algal Limestone Member, only 3 feet stratigraphically below an Early Ordovician fauna. Moreover, on Main Elk Creek, west of the map area, a bed of limestone containing an Early Ordovician fauna lies only 6½ feet above the Clinetop Algal Limestone Member. Inasmuch as Early Ordovician fossils are present so close above the Clinetop Member, the top of the member has been designated as the top of the Dotsero Formation (Bass and Northrop, 1953, p. 897). The upper boundary of the formation is thus readily distinguishable, because the Clinetop Member can be identified, wherever exposed, throughout the White River Plateau. The Dotsero Formation of Late Cambrian age contains considerable amounts of glauconite in many beds, whereas the overlying Manitou Formation of Early Ordovician age contains only minor amounts in a few beds. This condition is in accord with the observations of Cloud and Barnes (1948, p. 23), who state:

> Within the experience of the authors in the mid-Continent region, the presence of glauconite as a common accessory mineral is presumptive evidence of a Cambrian age for the rock in which it occurs.

The lower part of the Dotsero Formation commonly forms a brush-covered slope that recedes from the cliff of the underlying Sawatch Quartzite. The beds of the upper part of the formation are more resistant and form steeper slopes than the lower part; they commonly form ledges composed of several thin beds of limestone, conglomerate, and shale.

GLENWOOD CANYON MEMBER

The Glenwood Canyon Member includes all beds in the Dotsero Formation lying below a ledge-forming algal limestone that forms the top unit of the formation (Bass and Northrop, 1953, p. 897). The lower half of the member consists of thin beds of light-gray to tannish-gray dolomite, and a few thin beds of flat-pebble dolomite conglomerate interbedded with thin beds of light-greenish-gray very dolomitic shale. The upper half of the member consists of thin beds of flat-pebble limestone conglomerate and interbedded light-greenish-gray very limy shale. The preceding stratigraphic section was measured at the type locality of the member, in the SE¼ sec. 16, T. 5 S., R. 87 W.

The beds of conglomerate in the member are composed of flat pebbles of dense medium-gray limestone embedded in a gray limestone matrix; the pebbles range in length from one-fourth inch or less to 5 inches and rarely to 8 inches; many are 1 to 2 inches long; most are less than one-half inch thick; almost all have rounded edges. In most beds the flat pebbles lie at small but varying angles with the bedding planes; however, the steep inclination of some of the pebbles gives the rock a complex structure. Locally a few beds are composed of flat pebbles, most of which are oriented at very steep angles to the bedding planes. The limestone matrix between the pebbles weathers somewhat more readily than the pebbles, and this differential weathering produces a slightly irregular surface on the top and particularly on the edge of a bed. Most beds of conglomerate are 5 to 8 inches thick.

Most of the dolomite and limestone beds are 5 to 8 inches thick, but range from 2 inches to 1½ feet in thickness. The shale beds are generally much thinner; many are only 1 to 3 inches thick. Grains of glauconite are common throughout the member but are less abundant than in the dolomite beds of the underlying Sawatch Quartzite. Most beds are fine grained or dense. The few beds and lenses of limestone that contain the fossils are coarsely crystalline.

A dolomite bed 1 to 2 feet thick in the lower third of the member forms a conspicuous ledge in many places on the White River Plateau. The bed is composed of fairly dense light-tannish-gray dolomite; the hummocky or wavy upper surface of the bed suggests an algal origin. It was not determined whether the ledge-forming dolomite is the same bed at all places. Its position seems to range from 25 to 35 feet above the base of the formation. A similar bed, unit 18 of the previously described stratigraphic section, is 46 feet above the base of the formation.

The bases of almost all the dolomite, limestone, and conglomerate beds in the Dotsero Formation contain a latticework of stemlike casts that are referred to by some geologists as fucoid markings and by others as burrows. The beds of conglomerate commonly weather reddish; these beds, together with beds of conglomerate in the overlying Manitou Formation, constitute the "red-cast" beds of early reports.

Fossils were collected from the Glenwood Canyon Member at several localities. These localities are described and faunal lists are given in detail by Bass and Northrop (1953, p. 908–910). Identifications in the following composite list are by A. R. Palmer (*in* Bass and Northrop, 1953, p. 908–911).

Porifera:
    Sponge spicules
Graptozoa:
    Callograptus sp.
    Dendrograptus sp.
    See also list of forms cited by Bassett (1938, 1939)
Brachiopoda:
    Fragments of inarticulates
Trilobita:
    Bowmania cf. *B. americana* (Walcott)
    Briscoia sp.
    Dikelocephalus cf. *D. minnesotensis* Owen
    Plethometopus sp.
    Saukiella sp.
    Stenopilus sp.
    New genus A
    Unidentified fragment

According to Palmer these fossils are of Trempealeau (Late Cambrian) age.

CLINETOP ALGAL LIMESTONE MEMBER

The most distinctive unit in the Dotsero and Manitou Formations undifferentiated is the Clinetop Algal Limestone Member, a ledge-forming algal limestone and conglomerate unit, 3 to 5 feet thick, at the top of the Dotsero Formation.

At most places the lower half of the Clinetop Algal Limestone Member consists of coarse flat-pebble limestone conglomerate, and the upper half consists of crystalline to dense algal limestone with a crinkly to wavy structure, and some conglomerate. Almost everywhere that the member crops out on the White River Plateau it weathers to a bone-white ledge having a distinct lavender tint, particularly on its top surface. Aside from the lavender tint, the most distinctive features of the rock are the circular or disk-shaped swirl patterns developed on the top surface and the crinkled structure shown in vertical sections of the rock. The swirls range in diameter from 3 to 12 inches; many, from 4 to 8 inches. In the center of each swirl is a crudely circular or elliptical disk, 1 to 4 inches in diameter, which is a cross section of a cone-shaped core that is normal to the bedding and tapers downward through the rock; the cone is composed of limestone conglomerate and presumably is the material that was deposited between the closely spaced algal columns or stromatolites.

The alga represents an undescribed species of *Collenia*, according to J. Harlan Johnson (*in* Bass and Northrop, 1953, p. 900–901). Many pelmatozoan columnals, about one-sixteenth of an inch in diameter, and ring-shaped cross sections of an undescribed genus of sponge, ¾ to 1 inch in diameter, are commonly weathered into relief on the top surface of the bed. Trilobites are represented by *Eurekia* sp., and fragments of dikelocephalids; brachiopods are represented by undetermined fragments of articulates. According to Palmer (*in* Bass and Northrop, 1953, p. 896), the entire Dotsero Formation represents most of the Trempealeau stage—the uppermost stage of the Cambrian.

The Clinetop Algal Limestone Member is particularly well exposed and forms a conspicuous bone-white ledge in its type locality, where it trends northeastward across the NE¼ sec. 35, T. 3 S., R. 90 W.; in the SW¼ sec. 23, T. 3 S., R. 90 W.; and throughout a large park area in sec. 36, T. 3 S., R. 89 W., where the member is 3 feet 2 inches thick. The member forms a conspicuous lavender ledge in the SW¼ sec. 4, T. 4 S., R. 88 W. (unsurveyed). There, the upper, algal part of the member is 2 feet 9 inches thick, and the lower, conglomeratic part is concealed. The Clinetop Member persists northwestward from its type locality at least beyond the South Fork of the White River, for it is clearly exposed in the cliffs on South Fork in secs. 20 and 29, T. 2 S., R. 90 W.

The thickest development of the member observed is in the eastern part of Glenwood Canyon, in the SE¼ sec. 16, T. 5 S., R. 87 W., about 60 feet above U.S. Highway 6 on its north side, at the base of a prominent cliff. There a sequence of beds 10.5 feet thick, containing algal limestone, is 90 feet above the base of the Dotsero Formation. The beds in the sequence and shaly beds directly above and below it have an undulating or wavy structure; the algal limestone beds are hummocky.

In areas of gentle slopes the Clinetop Algal Limestone Member commonly forms a shoulder whose edge is marked by bare surfaces of the lavender-tinted bone-white rock. Locally, a few feet of interbedded flat-pebble limestone conglomerate and shale are exposed below it. In cliff faces, such as those in Glenwood Canyon, the member can be distinguished as a blocky unit having a wavy top and base.

Exceptional conditions must have prevailed in the White River Plateau region near the close of Cambrian time to permit the development of such an extensive algal limestone or biostrome as it was seen at many places throughout an area of 400 square miles. Clearly, the occurrence of flat-pebble conglomerates both below and above the biostrome suggests considerable agitation of the water in Late Cambrian time and again in Early Ordovician time. The widespread algal biostrome, however, may indicate a period of quiet water just prior to the close of the Cambrian.

MANITOU FORMATION

The Manitou Formation, as designated by Bass and Northrop (1953, p. 905), includes a little more than the upper half of the Dotsero dolomite of Bassett (1939, p. 1855–1858). The term Manitou Formation was used instead of Manitou Limestone, because in the White River Plateau the formation includes a variety of rock types. At most places the lower half of the formation (in Glenwood Canyon, where the formation is thickest, the lower three-fifths), the Dead Horse Conglomerate Member, consists chiefly of thin beds of gray flat-pebble limestone conglomerate interbedded with greenish-gray very calcareous shale and a few beds of limestone and dolomite that weather brown. The upper part of the formation, the Tie Gulch Dolomite Member, consists of regularly bedded, thin-bedded medium-brown somewhat siliceous dolomite that commonly forms a cliff. The thickness of the Manitou Formation in the four stratigraphic sections that were measured ranges from 80 feet on the South Fork of the White River to 155 feet in Glenwood Canyon.

DEAD HORSE CONGLOMERATE MEMBER

The lower part of the Manitou Formation, which consists largely of thin beds of gray flat-pebble limestone conglomerate, is called the Dead Horse Conglomerate Member (Bass and Northrop, 1953, p. 905). The member is exposed high in the cliffs along Dead Horse Creek, northwest of the northeast corner of sec. 30, T. 5 S., R. 87 W. A more accessible place is the spur on the north side of U.S. Highway 6, one-half mile northeast of the bridge over French Creek, where the member is clearly exposed. Many fossils were collected from the member at this locality, which was designated as the type section (Bass and Northrop, 1953, p. 905).

The beds of conglomerate in the member are similar to those in the Glenwood Canyon Member of the Dotsero Formation; the description of the conglomerate beds of that member given on page 11 of this report can be applied to the conglomerate beds in the Dead Horse Conglomerate Member. A few beds in the lowermost part of the member contain considerable glauconite, whereas a very few beds higher in the member contain only traces. Most beds of conglomerate are 5 to 8 inches thick but may range from 3 inches to a foot or more. The beds of shale are commonly thinner. The conglomerate weathers dull mottled gray and brown to brown, and many beds and slabs weather mottled red and brown; hence, the term "red-cast" has also been applied to these beds, as well as to those in the underlying Dotsero Formation. Sequences 5 to 20 feet thick of the interbedded conglomerate and shale commonly form alternating slopes and low ledges, rising above the shoulder of the underlying Clinetop Algal Limestone Member of the Dotsero Formation.

In 1951, 17 collections of Early Ordovician fossils, which include trilobites, brachiopods, gastropods, a cephalopod, conodonts, sponge spicules, pelmatozoans, and a graptolite, were made from the Dead Horse Conglomerate Member. Eight of these collections were made from Glenwood Canyon on the east side of a prominent spur in the cliffs north of U.S. Highway 6, one-half mile northeast of the bridge over French Creek, where the fossil-bearing beds are distributed through the sequence that extends from 3 to 93 feet above the base of the formation. The other nine fossil collections are from beds in the lower 35 feet of the formation at other places.

According to A. R. Palmer, who with others made the identifications, the fossils of the Dead Horse Conglomerate Member of the Manitou Formation are more diversified than those of the Dotsero Formation and suggest an early Beekmantown (Early Ordovician) age. Palmer has recognized two subzones in the Dead Horse Conglomerate

Member (Bass and Northrop, 1953, p. 906, 908–911). The basal subzone *A* yielded the following fossils:

Porifera:
    Sponge spicules
Graptozoa:
    Callograptus? sp.
    Fragments
Pelmatozoa:
    Fragments
Brachiopoda:
    Apheoorthis sp.
    Paleostrophia sp.
    Syntrophina sp.
    Fragments of inarticulates
Gastropoda:
    Fragments
Trilobita:
    Hystricurus? sp.
    Symphysurina sp. A
    S. sp. B
    S.? sp.
    New genus A?
    New genus B, aff. *Phoreotropis* Raymond
    Genus undet.
Conodonts:
    Cordylodus sp.
    Cyrtoniodus sp.
    C.? sp.
    Plectodina sp.
    Fragments

The upper subzone *B*, occurring in strata more than 60 feet above the algal bed in the Glenwood Canyon section, yielded the following fossils:

Porifera:
    Sponge spicules
Brachiopoda:
    Nanorthis sp.
    Syntrophina sp.
    Fragments of inarticulates
Gastropoda:
    Bucanella sp.
    Gasconadia sp.
    Three undet. genera
    Fragments
Cephalopoda:
    Burenoceras? sp.
Trilobita:
    Bellefontia sp.
    B.? cf. *B.? acuminiferentis* Ross
    Clelandia sp.
    Hystricurus sp.

Trilobita—Continued
    Symphysurina sp. C
    S. sp. undet.
Conodonts:
    Oistodus sp.
    Paltodus sp.
    Fragments

TIE GULCH DOLOMITE MEMBER

The upper part of the Manitou Formation consists of a regular and thin-bedded, medium-brown dolomite, commonly forming a cliff, which is called the Tie Gulch Dolomite Member (Bass and Northrop, 1953, p. 906). The member forms a cliff at the east end of the U.S. Highway 6 bridge over Tie Gulch, near the center of sec. 15, T. 5 S., R. 87 W. The exposures in the walls of Tie Gulch at this place constitute the type section of the member. The upper part of the cliffs at the type locality of the Dead Horse Conglomerate Member, one-half mile northeast of the mouth of French Creek, also contains excellent exposures of the Tie Gulch Dolomite Member. The thickness of the Tie Gulch Dolomite Member ranges from 47 feet in Glenwood Canyon to 66 feet on Main Elk Creek. On the South Fork of the White River the basal 27 feet of the dolomitic (upper) part of the Manitou Formation is composed chiefly of beds of flat-pebble conglomerate, which is overlain by thin-bedded dolomite similar to the rocks of the Tie Gulch Dolomite Member elsewhere. Most of the dolomite of the Tie Gulch Member is fine grained, but some is medium grained. Many beds are slightly siliceous, and a few are fairly sandy. Thin stringers of light-yellow chert are present on weathered surfaces in many places. On the South Fork of the White River a sequence of beds 7 feet thick at the base of the member is sandy and quartzitic. Traces of glauconite are present in only a few beds. No fossils were found in this member.

The Tie Gulch Dolomite Member constitutes one of the most readily recognizable units throughout the White River Plateau. In most canyons it forms a light-brown cliff that caps steep slopes, and its top forms a prominent bench where the basal part of the overlying Parting Member of the Chaffee Formation is eroded back from the edge of the cliff. In areas of grass-covered slopes, such as the area between the headwaters of Grizzly Creek and Deep Lake, the member forms a series of thin-bedded rock ledges 3 to 8 feet high, interspersed with grass-covered slopes, rising to the basal quartzite ledge of the overlying Chaffee Formation.

UPPER DEVONIAN

CHAFFEE FORMATION

The Chaffee Formation of Late Devonian age unconformably overlies the Manitou Formation. The bedding in the Chaffee Formation is generally parallel with that in the underlying Manitou.   The Chaffee consists of two members that are widely different in composition.   The Parting Member occupies about the lower one-third of the formation and consists of interbedded shale and quartzite.   The Dyer Member occupies the upper two-thirds of the formation and consists of interbedded limestone and dolomite.

The following section of the Chaffee Formation was measured in the cliff north of U.S. Highway 6 near the east end of Glenwood Canyon about 400 feet west of the Eagle-Garfield County line, and offset a few hundred feet west upward through the formation.

Stratigraphic section of the Chaffee Formation on the north side of U.S. Highway 6 and directly west of the Eagle-Garfield County line, near the center of the S½ sec. 11, T. 5 S., R. 87 W.

Leadville Limestone (Mississippian):

Very sandy dolomite equivalent to the Gilman Sandstone Member of the Leadville Limestone, which here forms a prominent ledge, is 19 ft 8 in. thick, and contains disconformable bedding in its lowermost 3½ ft.

Chaffee Formation (Upper Devonian):

Dyer Member:

		Ft	In
31.	Limestone and dolomite, interbedded, finely crystalline to dense, gray; contains a few nodules of gray to dark-gray chert	17	0
30.	Same as unit 31 without chert	18	0
29.	Same as unit 31	4	0
28.	Same as unit 30	14	0
27.	Limestone, finely crystalline to dense, gray	9	0
26.	Limestone breccia, dense, thin-bedded, gray	4	0
25.	Dolomite, cryptocrystalline, gray; weathers light brown or tan	2	0
24.	Dolomite, dense, gray, and interbedded gray shale	1	4
23.	Dolomite, sandy, dense, light-gray; rounded, frosted medium and fine grains of quartz sand	3	0
22.	Shale, calcareous, gray	1	8
21.	Limestone, very finely crystalline, massive, light-gray	4	0
20.	Dolomite, very finely crystalline, very light gray; weathers tan; has conchoidal fracture; rarely fossiliferous, but fossils were collected from a bed 7 ft above base. This is a distinctive brown unit that persists throughout the map area	26	0
19.	Limestone, finely crystalline, in part slightly argillaceous, very fossiliferous, dark-gray; forms a dull dark-gray ledge whose surface contains solution cavities 6 in. to more than a foot in diameter; weathers nodular, hence the field name of "knobbly bed"	48	0
	Thickness of Dyer Member	152	0

Chaffee Formation (Upper Devonian)—Continued

Parting Member:

		Ft	In
18.	Quartzite, dense, clear quartz grains, medium to coarse, weathers tan; in beds 2-3 ft thick; some beds are crossbedded; forms a conspicuous tan ledge; contains a few thin shale partings	20	6
17.	Interbedded limestone, thin-bedded, dark-gray, and thin-bedded dark-gray dolomite and sandy dolomite and black shale	12	0
16.	Covered; much is probably black shale	10	6
15.	Shale, dolomitic, sandy, nodular	5	6
14.	Quartzite, dense, glassy grains; weathers light tan; in 4- to 6-in. beds with thin shale partings	4	6
13.	Shale, dark-green	3	4
12.	Quartzite, dense, glassy quartz grains; interbedded with very thin beds of green shale; weathers tan; unit forms a ledge	8	0
11.	Shale, green, with thin lenses of quartzite		6
10.	Dolomite, finely crystalline to dense, gray, weathers tan	1	8
9.	Shale, blocky, dark-green to olive-green	2	5
8.	Dolomite, finely crystalline, dark-gray, weathers tan	1	8
7.	Dolomite, calcitic, shaly, dense, thin-bedded, greenish-gray	1	6
6.	Dolomite, very sandy, gray; weathers tan; forms a ledge	2	0
5.	Quartzite, clear, with lenses of green shale and black dolomite		6
4.	Shale and interbedded slightly calcitic dolomite, dense to finely crystalline, dark-gray to black; contains several thin lenses of dense glassy quartzite; shale is micaceous	6	0
3.	Dolomite, finely crystalline, dark-gray to black; weathers rusty tan	2	0
2.	Shale, sandy, silty, micaceous, dark-greenish-gray; contains thin lenses of dolomite; weathers rusty tan	6	5
1.	Quartzite, white; includes sandy shale lenses; includes pebbles in lower part	5	0
	Thickness of Parting Member	94	0
	Total thickness of Chaffee Formation	246	0

PARTING MEMBER

The thickness of the Parting Member ranges from 63 to 95 feet in 10 stratigraphic sections that were measured. The Parting Member consists of interbedded light-green shale, black shale, light-tan glassy quartzite, and some dolomite and sandy dolomite. It is the most variable unit in the sequence of rocks of Paleozoic age; stratigraphic sections only a few hundred feet apart show considerable variation, bed for bed. The beds of shale are commonly micaceous, many are sandy, and some contain salt-crystal casts. The beds of quartzite are lenticular and crossbedding at steep angles is common. The grain size commonly is variable, ranging from very fine to coarse, and, locally, to pebble size.

Small collections of fish remains were made from the Parting Member as follows:

Locality 1.—North wall of Glenwood Canyon, directly above U.S. Highway 6, in the S½ sec. 11, T. 5 S., R. 87 W., Garfield County, Colo. This locality is near the Eagle-Garfield County line at the northeast and upper end of Glenwood Canyon, about 12 miles northeast of Glenwood Springs. Fossils were collected in 1947 from shaly dolomite and also from quartzite. According to D. H. Dunkle (written communication, June 2, 1949),

These bone fragments, as can be demonstrated histologically, pertain to the archaic fish group known as the Antiarchi.

Inability to make out complete outlines of any one element prohibits a definite generic and specific identification.

Locality 2.—East wall of Dead Horse Creek Canyon, in the NE¼ sec. 12, T. 5 S., R. 88 W., Garfield County, Colo. This locality is a little less than 5 miles west of locality 1. A small lot was collected here by Bass on September 28, 1950. According to Dunkle (written communication, Jan. 19, 1951), this material represents the antiarch *Bothriolepis* cf. *B. coloradensis* Eastman. Present are

disassociated plates among which are recognized a centro-nuchal, postero-medio-dorsal, postero-ventro-lateral, medio-ventral, and a latero-marginal 2, according to the classification of bones employed by Stensiö, 1948, and Denison, 1951, among others.

Subsequently, another collection was made at this locality by Bass and Northrop on Aug. 14, 1951. Dunkle (written communication, Mar. 16, 1953) reported that

all materials in this lot appear to pertain to the antiarch *Bothriolepis coloradensis* Eastman.

Locality 3.—North wall of Deep Creek Canyon, about three-fourths of a mile upstream from the logging road, in either the NW. cor. sec. 25 or the SW. cor. sec. 24, T. 4 S., R. 87 W., Eagle County, Colo. This locality is about 4 miles north and one-half mile east of locality 1 and about 6 miles northeast of locality 2. Two small lots were collected here by Bass on August 21, 1951. One lot, from a bed 30 feet above the base of the Parting member, contains "scale and bone referred to an indeterminate holoptychiid and a fragmentary ?arthrodiran plate" (Dunkle, written communication, Mar. 16, 1953). The other lot, from a bed 41 feet above the base of the member, contains "indeterminate scales and bones of a holoptychiid fish."

A classification of these fossil fish is given below:

Phylum Chordata
   Subphylum Vertebrata
      Class Placodermi—Primitive armored fishes
         Order Antiarchi—Small fishes with jointed pectoral fins
            Family Asterolepidae
               Bothriolepis coloradensis Eastman
         Order Arthrodira—Armored fishes with jointed necks
      Class Osteichthyes—Bony fishes
         Subclass Choanichthyes—Air-breathing fishes
            Order Crossopterygii—Lobe-finned fishes
               Suborder Rhipidistia—Ancestors of the amphibians
                  Family Holoptychiidae

Dunkle (written communication, March 16, 1953) comments:

Bothriolepis coloradensis was originally described from the Elbert formation and antiarch remains referred to the same species have subsequently been reported from the base of the Temple Butte formation, Arizona, and the Parting member of the Chaffee formation near Glenwood Springs, Colorado. *Bothriolepis* is considered a primary fresh-water fish and had cosmopolitan distribution during the Upper Devonian. The crossopterygian fishes of the family Holoptychiidae are likewise primary fresh-water fishes and their remains display a cosmopolitan distribution with a geologic range from the Middle Devonian through the Upper Devonian. Bones and scales have been the subject of numerous reports from the Upper Devonian deposits of the Rocky Mountain area. Preservation of the present examples does not permit generic and specific identification. A variety of names, however, have been applied to such remains in the mentioned region: *Holoptychius* cf. *giganteus* Agassiz (Eastman, 1904, from the Elbert formation of La Plata County, Colorado); *Glyptopomus sayrei* and/or *Holoptychius* sp. (Bryant and Johnson, 1936, and Denison, 1951, Fossil Ridge, Gunnison County, Colorado); *Holoptychius* sp. (Gidley, *in* Schuchert, 1918, from the Temple Butte formation, Arizona); and *Litoptychus bryanti* Denison (1951, from the Chaffee formation, Gunnison County, Colorado).

The presumed arthrodiran plate could indicate a marine environment. More conclusively, however, the presence of marine elements in the Parting member of the Chaffee formation has been previously suggested (Bryant and Johnson, 1936) by the recognition of such forms as *Ctenacanthus*, *Sandalodus*, and other arthrodiran bones from a locality on Gribbles Creek, Fremont County, Colorado.

As implied by various references (Gidley, *in* Schuchert, 1918; Eastman, 1917; and Denison, 1951), *Bothriolepis coloradensis* is indistinguishable from *B. nitida* Newberry from the Chemung beds of Pennsylvania and New York. Heretofore, the preservation of available materials has not permitted proof of this suggested relationship. Among the present specimens from Locality 2 are several apparently well articulated examples which might help in this regard.

DYER MEMBER

The Dyer Member of the Chaffee Formation consists of fairly thick units of limestone and dolomite. A few beds are sandy and others are cherty. The thickness of the member ranges from 140 to 175 feet in five sections that were measured. The basal unit is a cliff-forming dark-gray limestone, ranging from 48 to 75 feet in thickness; it constitutes one of the main key units in the Paleozoic sequence. It can be identified readily in canyon outcrops by the nearly sheer gray wall that it forms, rising above the slope formed by the Parting Member, and by the presence of solution cavities. The rock weathers nodular, hence was referred to in the field as the "knobbly bed." It is abundantly fossiliferous at most places. A persistent unit, 10 to 30 feet or more thick, of relatively dense light-brown-weathering dolomite overlies the "knobbly bed." Interbedded dense gray dolomite and limestone, some beds of which contain rounded grains of quartz sand, constitute the upper half of the formation.

Parts of the Dyer Member have yielded an abundant and well-preserved fauna of more than 50 species, predominantly brachiopods (34 species) but including also a few gastropods, cephalopods, pelecypods, anthozoans, bryozoans, annelids, echinoderms, and fish. Collections from several localities were identified at different times by P. E. Cloud, Jr., G. A. Cooper, and S. A. Northrop, as shown in the following table:

TABLE 1.—*Fauna of the Dyer Member of the Chaffee Formation*

	Dyer Member of Chaffee Formation, Glenwood Springs area, Colorado								Ouray Limestone (restricted), southwestern Colorado	Percha Shale, southwestern New Mexico
	1	2	3	4	5	6	7	8	9	10
Anthozoa										
"Streptelasma" sp	×								—	—
"Zaphrentis" sp		×							—	×
Acrophyllid coral				×			×		—	—
Bryozoa										
Ptiloporella sp				×				×	—	—
Bryozoans, undet		×		×				×	—	—
Brachiopoda										
Athyris coloradensis Girty	×	×	×	×	×	×	×	×	×	×
transversa (Stainbrook)				×						×
Bispinoproductus varispinosus Stainbrook[1]	×									×
sp						×				
Camarotoechia sobrina Stainbrook	×	×		×	×		×		×	×
Composita? sp						×				
Cyrtiopsis animasensis (Girty)	×	×	×		×				×	×
sp. aff. C. animasensis (Girty)							×			
conicula (Girty)			×						×	—
kindlei (Stainbrook)	×			×				×		×
n. sp							×			

footnotes at end of table.

TABLE 1.—*Fauna of the Dyer Member of the Chaffee Formation*—Continued

	Dyer Member of Chaffee Formation, Glenwood Springs area, Colorado								Ouray Limestone (restricted), southwestern Colorado	Percha Shale, southwestern New Mexico
	1	2	3	4	5	6	7	8	9	10
Brachiopoda—Continued										
Echinoconchus? laminatus (Kindle)	×						×			×
Eumetria? sp								×		
"Eunella" sp	×								×	×
Heteralosia nupera Stainbrook [2]						×				×
Leioproductus coloradensis (Kindle)	×	×		×	×	×	×	×	×	×
Paurorhyncha endlichi (Meek)	×	×	×	×	×	×	×		×	?
cf. *P. cooperi* Stainbrook			×							×
sp								×		
Planoproductus depressus (Kindle)	×	×	×	×	×					
cf. *P. hillsboroensis* (Kindle)							×	×		×
Porostictia perchaensis (Stainbrook)				×						×
sp		×				×				
Pugnoides sp							×			
Rhynchospirina sp. aff. *R. scansa*						×				
? sp							×	×		
Schizophoria australis Kindle	×	×	×	×	×	×				×
cf. *S. australis* Kindle				×			×	×		
Schuchertella chemungensis (Conrad)	×								×	
sp		×		×	×		×			×
Strophopleura notabilis (Kindle)	×				×					×
sp				×			×			
Inarticulate brachiopod fragment		×								
Meristelloid brachiopod						×				
Pelecypoda										
aff. *Aviculopecten*, n. gen., n. sp						×				
Conocardium sp							×			
Pelecypod, gen. undet				×						
Gastropoda										
"Bellerophon" sp	×			×					×	
Isonema humile Meek	×		×		×				×	
Platyceras (*Platyostoma*) cf. *P. gigantea* (Hall and Whitfield)				×			×	×	×	
sp				×						
Straparolus (*Euomphalus*) cf. *S. eurekensis* Walcott	×									
sp				×	×		×			
High-spired gastropods, gen. and sp. undet				×			×	×		

See footnotes at end of table.

TABLE 1.—*Fauna of the Dyer Member of the Chaffee Formation*—Continued

	Dyer Member of Chaffee Formation, Glenwood Springs area, Colorado								Ouray Limestone (restricted), southwestern Colorado	Percha Shale, southwestern New Mexico
	1	2	3	4	5	6	7	8	9	10
Cephalopoda										
Coleolus sp				X					X	—
Spyroceras? sp							X		X	—
Gomphoceroid cephalopod, gen. and sp. undet		X		X					—	—
Nautiloids, gen. and sp. undet		X		X			X		—	—
Annelida										
Cornulites sp				X					—	—
Spirorbis sp. on *Planoproductus* cf. *P. depressus* (Kindle)								X	—	—
Echinoderma										
Pelmatozoan columnals				X			X	X	—	—
Echinodermal pinnules or brachioles							X		—	—
Vertebrata										
Fish tooth, fragments							X		—	—

¹=*Leioproductus varispinosus* (Stainbrook) (Muir-Wood and Cooper, 1960).
²=*Acanthatia nupera* (Stainbrook) (Muir-Wood and Cooper, 1960).

Localities at which fossils were collected from rocks of the Dyer Member of the Chaffee Formation

No.   *Collector, year of collection, stratigraphic assignment, description of locality, and person identifying*

1   E. M. Kindle, before 1909. Ouray Limestone. Glenwood Canyon of Grand [Colorado] River near Shoshone and near Glenwood Springs, Garfield County, Colo. E. M. Kindle (1909).

2   S. A. Northrop and N. W. Bass, 1947. Dyer Member of Chaffee Formation. North wall of Glenwood Canyon, directly above U.S. Highway 6, at Eagle County–Garfield County line; collected in both counties. This locality is about 12 miles northeast of Glenwood Springs. P. E. Cloud, Jr. (written communication, Nov. 15, 1949).

3   S. A. Northrop and N. W. Bass, 1948. Same as locality 2. G. A. Cooper (written communication, June 2, 1949).

4   S. A. Northrop, 1951. Same as locality 2. S. A. Northrop, emended by P. E. Cloud, Jr. (written communication, July 13, 1954).

5   S. A. Northrop and N. W. Bass, 1948. Dyer Member of Chaffee Formation. Cliff north of Gallagher's cabin, Garfield County, Colo.; this locality is on the White River Plateau, about 10 miles north of Glenwood Springs and 14 miles west of locality 2; Blue Lake itself is in sec. 27, T. 4 S., R. 89 W. G. A. Cooper (written communication, June 2, 1949).

6   S. A. Northrop and N. W. Bass, 1948. Dyer Member of Chaffee Formation. North wall of Sweetwater Creek Canyon near junction of Turret Creek, upstream from Sweetwater Lake; NE¼ sec. 8, T. 3 S., R. 87 W., Garfield County, Colo. This locality is about 13 miles north of locality 2 and 13 miles northeast of locality 5. G. A. Cooper (written communication, June 2, 1949).

7   S. A. Northrop, N. W. Bass, and G. M. Richmond, 1951. Same as locality 6. S. A. Northrop, emended by P. E. Cloud, Jr. (written communication, July 13, 1954).

No.	Collector, year of collection, stratigraphic assignment, description of locality, and person identifying
8	S. A. Northrop and N. W. Bass, 1951. Dyer Member of Chaffee Formation. Cottonwood Creek, south of Colorado River, sec. 13, T. 5 S., R. 87 W., Eagle County, Colo. This locality is a little more than 1 mile southeast of locality 2. S. A. Northrop, emended by P. E. Cloud, Jr. (written communication, July 13, 1954).
9	E. M. Kindle and others, prior to 1909. Ouray Limestone (restricted). Rockwood and other localities in southwestern Colorado. E. M. Kindle (1909).
10	E. M. Kindle and others, prior to 1909; M. A. Stainbrook, prior to 1947. Percha Shale. Kingston, Lake Valley, and Silver City, all in southwestern New Mexico. E. M. Kindle (1909); M. A. Stainbrook (1947).

Further collections were made by the authors in 1951. These were studied by Northrop and then sent to P. E. Cloud, Jr., for checking. Northrop's identifications were checked and emended by Cloud, in consultation with Helen Duncan and Ellis Yochelson. Cloud (written communication, July 13, 1954) reported as follows:

As to age, I would call this [material] high Devonian without much question in my own mind, but with due recognition of the fact that one could honestly argue for a Mississippian age. The big *Schizophoric*, the common *Cyrtiopsis*, and the *Paurorhyncha* imply correlation with the Percha, Pinyon Peak, "lower Ouray," and probably some part of the upper Threeforks. The big smooth productellids suggest Upper Devonian to me, as does the spiriferid *Strophopleura*. The gastropod subgenus *Platyostoma* is also a Devonian type. Helen Duncan says the bryozoan *Ptiloporella* is a common Devonian genus, though it ranges to the Permian. Also according to Helen Duncan the acrophyllid coral is of a type common in the Percha, but unknown to her from indisputable Mississippian. The *Echinoconchus*(?) and *Paraphorhynchus* [now *Porostictia*] would by themselves suggest Mississippian, but in association with the rest of the fauna I find it more logical to think of them as praecursor elements.

In his paper on the brachiopods of the Percha Shale of New Mexico and Arizona, Stainbrook (1947, p. 298–302) had summarized the conflicting evidence, concluding that the "evidence afforded by the brachiopods is preponderantly in favor of the Mississippian and appears convincing." In a paper titled "Discrimination of Late Upper Devonian," C. H. Crickmay (1952) discussed the general problem of the Devonian-Mississippian boundary in western North America. He concluded that a number of brachiopod species formerly assigned to *Cyrtospirifer*, such as *C. animasensis*, *C. kindlei*, *C. conicula*, and others, should be assigned to Grabau's Chinese genus *Cyrtiopsis*. *Cyrtiopsis* is stratigraphically younger than *Cyrtospirifer* and Crickmay believes that *Cyrtiopsis* is a reliable guide to the uppermost Devonian and that it does not extend into the Mississippian. Cooper (1954, p. 325) in a paper titled "Unusual Devonian Brachiopods" commented on the evidence of Devonian age provided by the goniatites and brachiopods that occur in the Percha Shale.

In a memorandum dated November 15, 1949, concerning collections from the Dyer Member submitted in 1947, P. E. Cloud, Jr., reported that the fauna found in the Dyer is correlative with that of the Percha Shale in New Mexico and Conewango Formation of New York. Cloud said that:

Paraphorhynchus [later assigned to a new genus, *Porostictia*] suggests Mississippian; *Cyrtospirifer* [later assigned to *Cyrtiopsis*] suggests Devonian; general aspect of fauna seems to me more like Devonian. Although this fauna can be closely placed in actuality, it is involved in a boundary line problem * * *. In official U.S. Geological Survey classification it is regarded as Upper Devonian.

In a memorandum concerning material submitted in 1948, G. A. Cooper (written communication, June 2, 1949) wrote that Stainbrook in 1947 put the fauna of the Percha Shale, a correlative of the Chaffee, in the Mississippian. He notes that *Rhynchospirina* like *R. scansa* occurs in strata probably of earliest Mississippian age (Cussewago age), but he believes this age change is debatable

because I have some evidence that the Percha is high Devonian (Conewango) in age. I feel the same about the Chaffee fauna, that it is really of high Devonian age, not Mississippian.

CARBONIFEROUS SYSTEMS

MISSISSIPPIAN

LEADVILLE LIMESTONE

The Leadville Limestone, whose thickness ranges from 175 to 225 feet, consists chiefly of limestone, although the lower one-third of the formation contains interbedded dolomite and limestone, many beds of which contain dark-gray chert. In the basal 20 to 30 feet the dolomite is very sandy and locally some beds of limestone are sandy. This lowermost portion of the formation is believed to be equivalent to the Gilman Sandstone Member, which is at the base of the Leadville dolomite near Minturn and Leadville (Tweto, 1949), 35 and 50 miles, respectively, southeast of the Glenwood Springs quadrangle. The bedding planes in this lower portion are wavy to irregular and probably represent minor disconformities. This unit was useful in identifying the boundary between the Leadville Limestone and the underlying Chaffee Formation, for the general lithology of the beds above and directly below it is similar and the beds are only sparingly fossiliferous.

The upper half of the Leadville consists of massive gray coarsely oolitic limestone. Oolites are present in the limestone throughout the map area and at many other places in northwestern Colorado. The Leadville Limestone forms the most striking outcrops of all the formations. Its light-gray cliff, capping steep slopes and cliffs, is conspicuous throughout much of the area. The surface formed on the Leadville Limestone in a broad flat area in secs. 15, 21 and 22, T. 4 S., R. 89 W., is marked by nearly straight branching channels, whose origin was not determined. The channel bottoms lie as much as 40 feet below the general surface on the Leadville; they are devoid of trees but are within a thickly forested area.

Following is a stratigraphic section of the Leadville Limestone measured near the east end of Glenwood Canyon.

Stratigraphic section of the Leadville Limestone on the north side of U.S. Highway 6, about 500 feet west of the Eagle-Garfield County line, in the S1/2 sec. 11, T. 5 S., R. 87 W.

Molas Formation (Pennsylvanian):
    Dull purplish-red clay capping high cliff of limestone.
Unconformity.
Leadville Limestone (Mississippian):

	Ft	In
6. Limestone, very oolitic, finely to coarsely crystalline, gray to light-gray, massive; forms the most conspicuous light-gray cliff in the area	101	6
5. Limestone and dolomite, interbedded, gray to dark-gray, finely crystalline to dense; contains a few nodules and lenses of gray and dark-gray chert commonly 2 to 4 in. thick and 4 to 18 in. long; weathers light gray, and with adjacent units forms cliff; generally massive but some beds up to 4 ft thick are slightly argillaceous and weather with a cleaty vertical fracture	38	6
4. Limestone, gray to dark-gray, finely crystalline, massive; includes a few nodules of dark-gray chert	8	0
3. Limestone, slightly argillaceous, very finely crystalline to dense; gray, weathers with a cleaty vertical fracture; forms a notch in the cliff	7	4
2. Dolomite, finely crystalline to dense, sandy, rounded medium quartz grains, generally massive beds with irregular to wavy bedding planes, gray; weathers light gray; forms a prominent shoulder and bench in steep slopes and cliffs	16	2
1. Dolomite, very sandy, rounded, medium to coarse quartz grains; irregular bedding planes, disconformable beds, gray; weathers light gray	3	6
(Items 1 and 2 constitute a unit believed to be equivalent to the Gilman Sandstone Member of the Leadville Dolomite near Leadville and Minturn.)		
Total thickness of Leadville Limestone	175	0

Determinable megafossils are rather uncommon in the Leadville Limestone of the map area and the writers did not attempt to do any systematic collecting from this formation. Many years ago Kindle (1909) cited a few fossils from near Shoshone and 30 years later Bassett (1939) cited a number of brachiopods from the east end of Glenwood Canyon. J. Harlan Johnson (1945) described and illustrated several new species of calcareous algae from the Leadville Limestone at Glenwood Springs. In the following composite list, unless otherwise indicated, species were determined by Johnson (194•

from the upper part of the Leadville at Glenwood Springs and at several other localities not far distant in Garfield, Eagle, and Pitkin Counties, Colo.

Algae:
  Coelosporella sp.
  Garwoodia sp. aff. *G. gregaria* (Nicholson)
      media Johnson (holotype)
  Girvanella? *nicholsoni* (Wethered)
  Gymnocodium sp.
  Ortonella coloradoensis Johnson (holotype)
      furcata Garwood
      cf. *O. kershopensis* Garwood
      sp.
  Solenopora glenwoodensis Johnson (holotype)
      similis Paul
  Spongiostroma? spp., including small colonies, perforating algae, algal coatings, and ooliths
Foraminifera:
  Endothyra spp.
  Small foraminifers, several genera
Porifera:
  Spicules
Anthozoa:
  Syringopora sp.
  Undetermined tetracorals
Bryozoa:
  Fragments, abundant and varied
Brachiopoda:
  Composita humilis [1]
  Dictyoclostus parviformis [1]
      "*semireticulatus*" [1]
  Dielasma? sp. [3]
  Eumetria verneuiliana (Hall) [3]
  Linoproductus ovatus (Hall) [1][3]
  Rhipidomella burlingtonensis (Hall) [1]
  Schellwienella inflata (White and Whitfield) [1]
  Spirifer centronatus Winchell [1]
  Torynifer cooperensis (Swallow) [1]
Gastropoda:
  Bellerophon sp.[3]
  Straparolus (*Euomphalus*) sp.[3]
  Small forms, undet.
Cephalopoda:
  Small forms
Ostracoda:
  Fragments

[1] Cited by C. F. Bassett (1939) from the upper part of the Leadville Limestone near the east end of Glenwood Canyon, sec. 15, T. 5 S., R. 87 W., Garfield County, Colo.
[3] Cited by E. M. Kindle (1909, p. 12); fossils collected by Kindle near Shoshone were determined by G. H. Girty.
[3] = *Ovatia ovata* (Hall) (Muir-Wood and Cooper, 1960).

Crinoidea:
    Fragments abundant
Echinoidea:
    Spines and a plate
Vertebrata:
    Fish teeth [4]

[4] Observed by Bass and Northrop, Aug. 25, 1948, near Crane Park, sec. 3, T. 4 S., R. 88 W. (unsurveyed), Garfield County, Colo.

Large productoids of the *Dictyoclostus* type occur in the uppermost part of the Leadville on Deep Creek. In addition, an abundance of large euomphalids and of 4-inch-long high-spired gastropods was observed in the upper part of the formation on East Brush Creek, southeast of Eagle, Eagle County, on July 26, 1951. An angular block of greenish chert from the Molas Formation on the promontory between the third and fourth gullies northeast of Glenwood Springs, south of the Colorado River and above the railroad tunnel, contains a single fragment of *Cleiothyridina?* sp. and numerous examples of *Spiriferina* cf. *S. solidirostris* (White). This rock presumably was derived from the Leadville Limestone.

Little is known about the stratigraphic ranges of the algae of Leadville age. Of the brachiopods, some are long-ranging forms, such as *Linoproductus ovatus,*[5] which ranges throughout the Mississippian. *Torynifer cooperensis* occurs in the Caballero Formation of Laudon and Bowsher, 1941 (of Early Mississippian Kinderhook age), and in the Lake Valley Limestone (of early Osage age) of southern New Mexico. On the other hand, *Eumetria verneuiliana* occurs in rocks of Osage through Chester age. Girty identified this species both from Glenwood Canyon and from Rockwood in southwestern Colorado. *Eumetria* cf. *E. verneuiliana* occurs in the Arroyo Penasco Formation of northern New Mexico; according to Gordon (*in* Fitzsimmons, Armstrong, and Gordon, 1956), this formation is of Meramec age.

The Leadville Limestone of the map area is lithologically similar, including the types of oolites, to the Leadville Limestone of the southern part of the Front Range in Colorado, which is designated by Maher (1950) to be of Meramec(?) age. It is not unlikely therefore that the upper part at least of the Leadville in the map area may be of Meramec age.

PENNSYLVANIAN

The Pennsylvanian System consists, in ascending order, of (a) a basal thin unit of purplish-red clay—the Molas Formation; (b) a thick sequence of dark-gray shale containing thin beds of fossiliferous

[5] *Ovatia ovata* (Muir-Wood and Cooper, 1960).

limestone, interbedded dark-gray shale, micaceous sandstone, and thick beds of gritstone—the Belden Formation; (c) a thick sequence of interbedded thick beds of gypsum and black shale—the Paradox Formation of Middle Pennsylvanian age; and (d) an unknown part of a thick sequence of red beds—the Maroon Formation of Pennsylvanian and Permian age. The total thickness of the Pennsylvanian rocks may exceed 4,000 feet in the western part of the area.

MOLAS FORMATION

Lying on a karst surface of the Leadville Limestone is a sequence of dull purplish-red clay that contains smooth nodules and boulders of chert and ranges in thickness from 1 to 25 feet. The sequence constitutes the Molas Formation. The only fossils found occur in chert derived from the Leadville Limestone. These include silicified concretions, 6 to 15 inches in diameter, each containing a large Syringopora or organ-pipe coral, almost certainly derived from the Leadville Limestone. The fauna from the uppermost part of the underlying Leadville is Osage to Meramec in age and beds that overlie the Molas Formation are of Morrow age, according to determinations of fossils by L. G. Henbest (Thomas, McCann, and Raman, 1945; Henbest, 1958) and by S. A. Northrop and T. G. Roberts for this report. The age of the Molas Formation of this area could, accordingly, range from Late Mississippian to Early Pennsylvanian.

It should be noted that Moore and others (1944, chart, col. 42) did not regard the Molas Formation of southwestern Colorado as older than Atoka age. Wengerd and Strickland (1954, p. 2167) extended the Molas as far down as middle Morrow age. According to Merrill and Winar (1958, p. 2109), the basal part of the Molas of southwestern Colorado may be as old as Meramec (Late Mississippian). Northrop believes that the Molas Formation of the Glenwood Springs area may be as old as Meramec. However, the Pennsylvanian age assigned by Girty (Girty, 1903; Cross and others, 1905) is retained in this report. In a paper on the significance of karst terrane and residuum in Upper Mississippian and Lower Pennsylvanian rocks of the Rocky Mountain region, Henbest (1958, p. 37) observed that although the Molas Formation is currently classed as Pennsylvanian in age by the Geological Survey, in his opinion, the Molas Formation "and correlative units * * * are Mississippian and Pennsylvanian in age and the parts belonging to each system can not be differentiated."

Following is a section of the Molas and Belden Formations measured on Deep Creek.

Stratigraphic section of Belden and Molas Formations on north slope of Deep Creek 1.7 miles up the Deep Creek road from its junction with the Colorado River road, in secs. 24 and 25, T. 4 S., R. 87 W.

Gypsum at base of Paradox Formation (Middle Pennsylvanian): near top of Onion Ridge.

	Ft	In
Belden Formation (Pennsylvanian):		
64. Shale, fissile, black	7	0
63. Conglomerate, crossbedded; much coarse angular quartz sandstone; some pebbles of feldspar, chert, and limestone; many pebbles range from ¼ to ½ in. in diameter; a few subrounded pebbles of chert as much as 2 in. in diameter; a few subrounded pebbles of limestone as much as 2 in. long; forms a prominent light-tan to gray ledge just below the crest of Onion Ridge	43	0
62. Shale, fissile, black, with a few beds of black dense limestone 1–8 in. thick, and a few zones of black dense mudstone concretions	58	0
61. Conglomerate, crossbedded; many coarse grains of quartz, with pebbles (some chert, many quartz) ranging from ¼ to 2½ in. in diameter; forms a prominent brown ledge between the top two big ledge formers of the formation (units 63 and 56)	10	0
60. Shale, fissile, black, with a few mudstone beds 2 in. thick	6	9
59. Sandstone, fine, silty, micaceous, greenish-tan; some interbedded silty shale; 4-ft bed of conglomerate, 3 ft above base	23	0
58. Shale, fissile, dark-gray to black; includes several black mudstone beds each 6 in. or less thick	15	0
57. Sandstone, micaceous, greenish-tan, interbedded with conglomerate of coarse quartz grains and pebbles as much as one-fourth in. in diameter	10	0
56. Conglomerate, crossbedded, micaceous, arkosic, contains many coarse quartz grains and ¼-in. pebbles; many pebbles 1 in. in diameter and a few larger; some chert and limestone; forms the most prominent light-gray ledge in the slope	24	0
55. Shale, silty, micaceous, black and dark-brownish-gray	3	6
54. Conglomerate and some sandstone (similar to unit 56)	24	0
53. Shale, fissile, black, upper half silty	5	7
52. Sandstone, fine, silty, micaceous, and some silty shale	10	0
51. Conglomerate, with torrential crossbedding; mostly coarse and very coarse quartz grains; some pebbles 1 in. and some clay balls 2½ in. in diameter; feldspar pebbles 1 in. long	6	0
50. Shale, fissile, black	7	6
49. Shale, black and dark-gray, and interbedded micaceous, greenish-tan siltstone	19	6
48. Sandstone and conglomerate, light-brown	7	8
47. Shale, fissile, black	5	1
46. Conglomerate and sandstone, micaceous, crossbedded; arkosic; forms a light-brown ledge	9	10

Belden Formation (Pennsylvanian)—Continued

		Ft	*In*
45.	Shale, black, containing nodules and lenses of ferruginous mudstone, a few of which are calcareous	12	8
44.	Sandstone and conglomerate, micaceous, in part calcareous	6	0
43.	Shale, fissile, black; contains several 1- to 3-in. beds of black mudstone, some of which have a hummocky surface and may be algal; 3 in. of micaceous silty shale near middle of unit	30	0
42.	Shale, black and dark-gray, interbedded with thin- to slabby-bedded micaceous greenish-tan siltstone and a little fine-grained to very fine grained micaceous greenish-tan sandstone; a 1-ft bed of dark-gray earthy limestone is 5 ft below the top	27	0
41.	Shale, black to dark-gray; a very few thin beds of greenish-tan micaceous siltstone; a few thin beds of calcareous dark-gray mudstone; a very few thin hummocky beds of limestone, which may be algal; a 1-ft bed of fossiliferous gray limestone 17 ft below the top	65	0
40.	Shale, black; includes a few beds of calcareous, micaceous greenish-gray siltstone, and a few beds, 6–12 in. thick, of dark-gray dense limestone, some of which are algal	54	6
39.	Shale, black; includes several beds, 3–9 in. thick, of dark-gray dense limestone, which contain fecal(?) pellets; several beds are algal; a prominent algal limestone bed is 6 ft above the base	44	0
38.	Limestone, dark-gray, finely crystalline, fossiliferous, weathers thin bedded; middle part includes thin beds of very calcareous dark-gray shale; basal bed of limestone, 1 ft 2 in. thick, has crenulated structure, probably algal	5	0
37.	Shale, silty, micaceous, dark-gray	1	6
36.	Shale, dark-gray; includes near the middle a 5-in. bed of dense dark-gray limestone	4	6
35.	Limestone, dense, dark-gray	1	5
34.	Mostly covered; probably black shale	11	0
33.	Mudstone, limy, contains dark-gray limestone nodules	1	4
32.	Shale, fissile, noncalcareous, in part finely micaceous, dark-gray to black; 6 ft 7 in. below top includes a bed, 1 ft 2 in. thick, of fine micaceous greenish-gray thinly laminated sandstone; 11 ft 1 in. below top a bed, 9 in. thick, of sandy dark-gray limestone with abundant dark-gray fecal(?) pellets ⅛ in. or less long; and 4 ft 5 in. above base, a micaceous calcareous greenish-tan thin-bedded unit, 1 ft 6 in. thick, of interbedded sandstone, siltstone, and shale	22	3
31.	Mudstone and nodular limestone with abundant fecal(?) pellets ⅛ in. long, gray and dark-gray	1	9
30.	Shale, calcareous, dark-gray, weathers light tan; includes 3 ft 4 in. below top a 6-in. bed of dark-gray mudstone; and 5 ft 3 in. below top a 1-ft 3-in. bed of nodular dark-gray limestone	8	4

Belden Formation (Pennsylvanian)—Continued

	Ft	In
29. Limestone, in two beds separated by a bed of clay 10 in. thick; all dark gray and fossiliferous; upper limestone, 1 ft 3 in. thick, is nodular, and lower limestone, 5 in. thick, is dense	2	6
28. Shale and clay, in part calcareous, dark-gray; lower third includes beds of dark-gray mudstone each 2–3 in. thick, some of which are calcareous; basal 1 ft 9 in. of unit is very calcareous, dark-gray shale	9	3
27. Limestone and shale, interbedded in thin beds; one bed of shaly limestone 2 ft 5 in. thick is dark gray and fossiliferous; several of the shale beds are noncalcareous	8	6
26. Shale and mudstone, dark-gray, in part calcareous	5	6
25. Limestone, in part argillaceous, medium- to dark-gray, weathers chunky and shaly	4	0
24. Shale, dark-gray, and 3 beds of limestone ranging from 4 in. to 1 ft 5 in. in thickness, in part fossiliferous; includes some thin beds of calcareous black mudstone; includes 1 ft above the base a 1-ft bed of ocherous-weathering clay, possibly bentonitic, that is a distinctive marker	11	6
23. Limestone; top 2 ft 4 in. is dark-gray dense limestone that weathers into slope above the underlying ledge; the next 2 ft 3 in. is recrystallized, spongy limestone with the lower 9 in. crenulated to wavy bedded, probably algal; the basal 6 ft 4 in. is medium-crystalline gray fossiliferous limestone, containing a little gray chert in nodules, that forms a prominent ledge	10	6
22. Shale, dark-gray, in part calcareous; includes several beds 8 in. to 5 ft 6 in. thick of gray mudstone, in part calcareous, that weather blocky; includes 2–3 beds 6–15 in. thick of gray dense limestone; includes in the upper half two units, each 6 ft thick, that are concealed	37	8
21. Limestone, with a few interbedded thin beds of calcareous shale; limestone is dark gray, finely crystalline to dense, fossiliferous, weathers into chips; forms a bedded ledge	8	9
20. Shale and mudstone, dark-gray; about half is concealed	10	0
19. Limestone and interbedded thin calcareous shale; limestone in upper half somewhat shaly and fossiliferous; limestone in lowermost 4 ft 8 in. is dark gray, dense, fossiliferous, and forms a jagged ledge	12	0
18. Shale, dark-gray, and interbedded thin beds of dark-gray limestone; top 3 ft is calcareous mudstone containing nodules of gray limestone	8	2
17. Limestone, with a few very thin beds of calcareous shale; limestone is argillaceous, fossiliferous, thin bedded to shelly; forms a thin-bedded ledge	7	0
16. Shale, dark-gray to black, fissile in lower part, chunky to massive in upper part; in part calcareous; includes a 1-ft bed of argillaceous gray to brown limestone in middle and nodules of gray limestone in basal 2½ ft	16	8
15. Limestone, argillaceous, finely crystalline to dense, dark-gray; and interbedded massive dull drab-gray shale	5	4

Belden Formation (Pennsylvanian)—Continued

	Ft	In
14. Limestone, fossiliferous, in 2-4 beds, finely crystalline, dark-gray, weathers blocky	5	0
13. Shale, dark-gray, and interbedded dark-gray limestone in 4-in. beds	5	0
12. Limestone, finely crystalline, medium-dark-gray; weathers blocky	2	3
11. Covered in part; considerable shale exposed, dark-gray in thin beds; a few 4-in. beds of dark-gray mudstone and a few beds 1 in. or more thick of dark-gray limestone	18	2
10. Limestone, gray, recrystallized, and interbedded dark-gray shale	6	0
9. Limestone, finely crystalline, argillaceous, fossiliferous; weathers thin bedded; forms a ledge	4	10
8. Shale, dark-gray, and interbedded mudstone in beds 1 to 2½ ft thick; about 25 percent covered	22	2
7. Limestone, finely crystalline, gray; includes a few partings of gray shale; basal 2 ft is very shaly	11	0
6. Mudstone, dark-gray, and interbedded dark-gray shale; includes 2 beds of gray limestone 5 ft apart and 1-2 ft thick; the upper limestone contains a trace of black chert; about 20 percent of the unit concealed	25	0
5. Limestone and interbedded limy shale; limestone is argillaceous, dense, and gray to dark gray; lower third of unit might be classed as limy mudstone; shale beds are very limy and gray to dark gray	25	0
4. Shale, black, and interbedded thin beds of dense very dark gray limestone with abundant ostracodes; includes a 1-ft coaly shale bed 18 ft above base; and a bed of gray mudstone, 4 ft 2 in. thick, 5 ft above the base; the basal 5 ft is mostly concealed	33	0
3. Sandstone, coarse, poorly sorted, micaceous, brown	8	0
2. Shale, dark-gray	1	7
1. Limestone, dense to crystalline, fossiliferous, medium-dark-gray, in beds 1-6 in. thick; weathers somewhat nodular	8	0
Total thickness of Belden Formation	924	0

Molas Formation (Pennsylvanian):

	Ft	In
Clay, deep-purplish-maroon to yellow and orange where weathered; contains smooth spherical to elliptical concretions or boulders of dark-brown quartzite as much as 10 in. in diameter. Rests on a karst surface on the underlying Leadville Limestone	9	0

Leadville Limestone (Mississippian):

Limestone, oolitic and coarsely crystalline, massive, gray, weathers light gray.

BELDEN FORMATION

The Belden Formation as the term is used herein is redefined from the definition by Brill (1944, p. 624) to extend from its contact with the underlying Molas Formation upward to the base of the lowest prominent gypsum bed of the overlying Paradox Formation. The thickness of the formation ranges from 600 feet at Glenwood Springs to 1,000 feet on Deep and Sweetwater Creeks. The Belden Formation, as redefined, is well exposed on the north slope of Deep Creek, a little less than 2 miles above its mouth. There the lower 675 feet of the formation consists of dark-gray to almost black shale and limy shale and interbedded thin-bedded gray, fossiliferous limestone, and a few very thin beds of black coaly shale. On Sweetwater Creek, which is 5 miles north of Deep Creek, a bed of coal 9 inches thick is present 28 feet above the base of the formation, and three beds of gypsum are present in the interval 87 to 147 feet above the base of the formation, as follows: a 3-foot 6-inch bed at 87 feet, a 12-foot bed at 100 feet, and a 4-foot bed at 143 feet. In another section 2½ miles downstream, there are four beds of gypsum in the interval 74 to 178 feet above the base of the formation, as follows: a 4-foot 8-inch bed at 74 feet. a 5-foot 10-inch bed at 106 feet, a 2-foot 4-inch bed at 154 feet, and a 6-foot 3-inch bed at 172 feet.

The uppermost 175 feet of this 675-foot sequence contains mostly shale and a few thin beds of fine-grained very micaceous greenish-tan sandstone. The sequence, 140 feet thick, lying next above the 675-foot sequence consists almost entirely of fissile dark-gray shale and thin-bedded, micaceous, greenish-tan sandstone. Above this are thick beds of coarse arkosic gritstone interbedded with arkosic conglomerate and fissile black shale. The beds of gritstone and conglomerate are extremely lenticular; many are from 10 to 20 feet thick; one is locally 50 feet thick.

Brill (1942, p. 1385–1387) proposed the name "Belden Shale Member" of the Battle Mountain Formation for strata previously called the "Weber Shale" in the Gore area east of Glenwood Springs. He cited a few fossils and stated that the age was Des Moines. In a later report on the late Paleozoic stratigraphy of west-central and northwestern Colorado, Brill (1944, p. 626) cited 23 species of fossils from the Belden Shale, which he then raised to formational rank, and correlated the Belden with the Cherokee Shale (Des Moines Series) of the Midcontinent region. In 1945, Henbest (in Thomas, McCann, and Raman, 1945) concluded that Brill's Belden Shale (as used in 1944) exposed at Glenwood Springs might be of Morrow age. M. L. Thompson (1945, p. 22, 43) concluded that the Belden Formation on Sweetwater Creek was Morrow in age. Eventually, Brill (1952, p. 814) suggested that "the Belden seems to be a facies that crosses time lines. It seems to be Morrowan, it is probably also Atokan, and may be Desmoinesian."

In the authors' opinion the lower 600 feet, more or less, of the Belden Formation is of Morrow age and the upper part of the formation is of Atoka age.

Fossils are locally abundant and diversified. A total of 114 lots were collected and the fauna (including a number of calcareous algae) numbers 258 species.[1] (See table 2 for a list of these fossils.) Large collections were made on Deep Creek from the lower 600 feet and on Sweetwater Creek from the lower 900 feet of the Belden Formation. Large collections were made also east of Glenwood Springs and south of Colorado River; smaller collections were made northwest of Glenwood Springs on Boiler Creek and on East Elk Creek.

Credit for identification of the Pennsylvanian fossils is as follows: Foraminifera, Lloyd G. Henbest and Thomas G. Roberts; Anthozoa and Bryozoa, Helen Duncan; Gastropoda, J. Brookes Knight; Ostracoda, I. G. Sohn; Crinoidea, Arthur L. Bowsher; and Vertebrata, David H. Dunkle. The junior author has identified the genera and species of other classes and, except where others are directly quoted, has supplied the age designations and discussions. Most of the paleontologic work was done during the period 1950–52 and reflects the taxonomy in use at that time. Changes that have been made in the generic assignment of a few brachiopod species since that time are incorporated here.

TABLE 2.—*Fossils of the Belden and Paradox Formations of the Glenwood Springs, Colorado, area*

	Belden Formation	Paradox Formation			Belden Formation	Paradox Formation
Algae						
Artophycus cf. A. columnaris Johnson	X			Algae, cauliflower type	X	
? n. sp.	X	-----		crusts resembling shells	-----	X
Calyptophycus? sp	X	-----		nodules, type 1	X	-----
Cryptozoon coloradensis Johnson	X	-----		type 2	X	-----
sp.	X	X		pellets	-----	X
Gouldina magna Johnson	X	-----		warty incrustations	-----	X
Leptophycus sp	-----	X		Algae (?), black rods	X	X
Shermarophycus gouldi Johnson	X	-----		oolites and (or) pisolites	X	-----
Stylophycus calcarius Johnson	X	-----				
cf. S. calcarius Johnson	X	-----				
Foraminifera						
Ammobaculites sp	X	-----		Paramillerella advena (Thompson)[1]	X	-----
Ammodiscus sp	X	-----		P. circuli (Thompson)[1]	X	-----
Bradyina sp. (early form)	X	-----		Profusulinella? sp	X	-----
Calcitornella sp	X	-----		Tetrataxis sp	X	-----
Climacammina sp	X	-----		Textularia? sp	X	-----
? sp	X	-----		Trepeilopsis? sp	X	-----
Cornuspira sp	X	-----		Ammodiscids	X	-----
Endothyra sp	X	-----		Calcitornellids	-----	X
? sp	X	-----		Calcitornellids (?)	X	-----
Endothyranella powersi (Harlton)	X	-----		Endothyrid (?)	X	-----
Glomospira sp	X	-----		Tolypamminids	X	X
? sp	X	-----				
Millerella inflecta Thompson (types)	X	-----				
cf. M. pressa Thompson[1]	X	-----				
sp. A[1]	X	-----				
sp. indet[1]	X	-----				
? sp	X	-----				

See footnotes at end of table.

[1] A detailed report on the Pennsylvanian fossils is being prepared by the junior author. For this reason, collecting localities are not set forth in the present report.

TABLE 2.—*Fossils of the Belden and Paradox Formations of the Glenwood Springs, Colorado, area*—Continued

	Belden Formation	Paradox Formation		Belden Formation	Paradox Formation
Porifera (?)					
Sponge spicules (?)		×			
Anthozoa					
Amplexocarinia? sp. indet	×		*Stereostylus* sp., aff. *S. adelus* Jeffords		×
Caninia sp	×		sp., aff. *S. lenis* Jeffords		×
Dibunophyllum? sp. indet	×		"*Zaphrentis*" cf. *Z. gibsoni* White of Girty	×	
Lophophyllidium cf. *L. ignotum* Moore and Jeffords		×			
cf. *L. profundum* (Milne Edwards and Haime)		×			
cf. *L. wewokanum* Jeffords		×			
Bryozoa					
Ascopora? sp. indet	×		*Prismopora*, probably *P. triangulata* White	×	
Cheilotrypa? sp. indet	×	×	sp	×	×
Cystodictya sp. indet	×		*Ramiporalia* sp. undet	×	
sp. indet., another species	×		? sp	×	
? sp. indet	×		*Rhabdomeson?* sp. indet		×
Dictyocladia sp. indet	×		*Rhombopora* sp. indet	×	
? sp. indet	×		sp. undet., additional species	×	
Eridopora? sp		×	*Tabulipora* sp. indet	×	
? sp. indet	×		? sp. undet	×	
Fenestella sp	×	×	Fenestellids, undet	×	×
sp. undet., additional species	×		Fenestelloid, gen. indet	×	
Fistulipora sp. undet	×		Fistuliporoid, gen. indet	×	×
Meekopora sp. indet	×		Rhomboporoids, gen. indet	×	×
Penniretepora sp. indet	×		Stenoporoid, gen. indet	×	
? sp		×			
Polypora sp. undet	×	×			
? sp. indet		×			
Brachiopoda					
Antiquatonia coloradoensis (Girty)	×	×	*Composita*—Continued		
cf. *A. coloradoensis* (Girty)	×		cf. *C. subtilita* (Hall)	×	×
cf. *A. hermosana* (Girty)		×	sp. A	×	
cf. *A. portlockiana crassicostata* (Dunbar and Condra)	×		sp. B	×	
? sp	×	×	sp	×	×
Beecheria cf. *B. arkansana* (Weller)	×		? sp	×	
Chonetes cf. *C. granulifer* Owen	×		*Derbyia crassa* (Meek and Hayden)	×	×
? sp		×	cf. *D. crassa* (Meek and Hayden)	×	
Chonetinella flemingi (Norwood and Pratten)	×		sp	×	×
cf. *C. flemingi* (Norwood and Pratten)	×		? sp	×	
flemingi crassiradiata (Dunbar and Condra)	×		*Echinoconchus?* sp [1]	×	
n. sp., aff. *C. flemingi crassiradiata* (Dunbar and Condra)	×		*Heteralosia?* sp	×	
? n. sp		×	*Horridonia bullata* (Mather)?	×	
? sp	×		*Hustedia miseri* Mather	×	
Cleiothyridina orbicularis (McChesney)	×	×	*mormoni* (Marcou)	×	×
Composita argentea (Shepard)	×		cf. *H. mormoni* (Marcou)		×
cf. *C. argentea* (Shepard)		×	? sp	×	
cf. *C. deflecta* Mather	×		*Juresania nebrascensis* (Owen)	×	
ovata Mather	×	×	*Lingula* n. sp. A, aff. *L. tighti* Herrick	×	
cf. *C. ovata* Mather	×		n. sp. B	×	
ozarkana Mather	×		sp. undet	×	×
cf. *C. ozarkana* Mather	×		? sp., indet	×	
subtilita (Hall)	×		*Linoproductus prattenianus* (Norwood and Pratten	×	×
			cf. *L. prattenianus* (Norwood and Pratten)	×	×
			sp	×	×
			? n. sp		×
			? sp	×	×

See footnotes at end of table.

TABLE 2.—*Fossils of the Belden and Paradox Formations of the Glenwood Springs, Colorado, area*—Continued

Brachiopoda—Continued

	Belden Formation	Paradox Formation		Belden Formation	Paradox Formation
Lissochonetes cf. L. geinitzianus (Waagen)	X		Punctospirifer kentuckiensis (Shumard)	X	X
Marginifera ingrata Girty	X		Rhipidomella carbonaria (Swallow)	X	
cf. M. muricatina Dunbar and Condra[2]	X		sp	X	
wabashensis (Norwood and Pratten)[4]		X	? sp	X	X
Mesolobus mesolobus decipiens (Girty)		X	Schizophoria? sp	X	
Neospirifer cf. N. cameratus (Morton)		X	Schuchertella sp	X	
dunbari R. H. King		X	? sp	X	
? sp		X	Spirifer occidentalis Girty	X	
Orbiculoidea missouriensis (Shumard)	X	X	cf. S. occidentalis Girty	X	X
sp.	X	X	cf. S. opimus Hall	X	
? sp	X		rockymontanus Marcou		X
Orthotichia schuchertensis Girty (holotype)	X		cf. S. rockymontanus Marcou	X	
Paeckelmannia n. sp., aff. P. derelicta R. H. King	X		Streptorhynchus affine Girty	X	
? sp	X		? sp	X	
Petrocrania modesta (White and St. John)	X	X	Trigonoglossa sp	X	
Phricodothyris perplexa (McChesney)		X	? sp	X	
cf. P. perplexa (McChesney)	X	X	Wellerella tetrahedra Dunbar and Condra		X
? sp	X		? sp		X
			Brachiopoda, gen. indet	X	X

Pelecypoda

	Belden Formation	Paradox Formation		Belden Formation	Paradox Formation
Acanthopecten carboniferus (Stevens)	X		Nucula cf. N. subrotundata Girty		X
Annuliconcha interlineata (Meek and Worthen)	X		? sp	X	X
interlineata (Meek and Worthen)?		X	Nuculopsis (Palaeonucula) n. sp., aff. N. (P.) croneisi Schenck	X	
? n. sp.	X		Palaeoneilo? sp	X	
Astartella varica McChesney?	X		Parallelodon tenuistriatus (Meek and Worthen)	X	
Aviculopecten sp. A		X	cf. P. tenuistriatus (Meek and Worthen)		X
sp. B, C, etc., additional species		X	Pernopecten cf. P. attenuatus (Herrick)	X	
sp. indet	X		Pleurophorus occidentalis Meek and Hayden	X	
? sp. indet	X	X	? sp	X	X
Aviculopinna peracuta (Shumard)		X	Pseudomonotis n. sp., aff. P. equistriata Beede	X	
sp	X	X	sp	X	
Bakewellia parva Meek and Hayden	X	X	? sp. undet	X	X
cf. B. parva Meek and Hayden	X		Schizodus subcircularis Herrick	X	
? sp. undet	X		sp	X	
Clinopistha sp	X		? sp		X
Culunana bellistriata (Stevens)	X		Septimyalina orthonota (Mather)	X	
? sp	X		cf. S. orthonota (Mather)	X	
Cypricardinia? sp	X		sinuosa (Morningstar)	X	
Edmondia ovata Meek and Worthen	X		sp	X	
cf. E. ovata Meek and Worthen	X		? sp	X	
Euchondria smithwickensis Newell	X		Solenomya ? sp	X	X
? sp	X		Streblochondria hertzeri (Meek)	X	
Limipecten n. sp., aff. L. koninckii (Meek and Worthen)	X		Yoldia sp	X	
sp	X		Pelecypoda, gen. indet	X	X
Myalina (Myalinella) cuneiformis Gurley	X	X			
(M.) cf. M. (M.) cuneiformis Gurley		X			
(M.) cuneiformis Gurley, n. var	X				
sp. indet	X				
? sp	X				

See footnotes at end of table.

TABLE 2.—*Fossils of the Belden and Paradox Formations of the Glenwood Springs, Colorado, area*—Continued

	Belden Formation	Paradox Formation		Belden Formation	Paradox Formation
Gastropoda					
Ananias? sp	X	--------	Retispira cf. R. tenuilineata (Gurley)	X	
Anematina sp	X	--------	cf. R. textiliformis (Gurley)	X	
Baylea? sp	X	--------	sp	X	
Donaldina stevensana (Meek and Worthen)? of Girty	X	--------	Shansiella cf. S. carbonaria (Norwood and Pratten) of Girty [5]	X	
sp	X	--------	Stegocoelia (Hypergonia) sp	X	
Euomphalus sp	X	--------	Strobeus cf. S. paludinaeformis (Hall) of Girty [5]	X	
Girtyspira? sp	X	--------	sp	X	
Glabrocingulum sp., aff. G. grayvillense (Norwood and Pratten) of Girty	X	--------	"Strophostylus cf. S. nanus" Meek and Worthen (of Girty) [5]	X	
Goniasma sp	X	--------	Worthenia? sp., aff. W. nebraskensis (Geinitz) of Girty [7]	X	
"Loxonema parvum" Cox (? of Girty) [5]	X	--------	Bellerophontids, gen. indet	X	
Naticopsis sp	X	--------	Murchisonid, n. gen. A	X	
Platyceras (Orthonychia) cf. P. (O.) parvum (Swallow)	X	--------	n. gen. B	X	
(O.) sp	X	--------	Gastropoda, gen. indet	X	X
Scaphopoda					
Plagioglypta sp	X	--------			
Cephalopoda					
Nautiloids, gen. indet	X	X	Goniatite, gen. indet	X	
Annelida					
Spirorbis sp. A	X	X	Serpuloid worm tubes in spreading masses, apparently identical with Monilipora prosseri (Beede) of Girty [5]	X	
sp. B	X	--------			
cf. S. sp. B	X	X			
? sp	X	X			
Trilobita					
Ditomopyge sp	X	X	Trilobita, gen. undet., fragments	X	X
Sevillia trinucleata (Herrick)	X	X			
Ostracoda					
Bairdiacypris cf. B. punctata Scott	X	--------	Paraparchites? sp. indet. 1	X	--------
"Beyrichia" sp. of Girty	X	--------	? sp. indet. 2	X	--------
Casellina n. sp	X	--------	? sp. indet. 3	X	--------
Jonesina? sp., aff. J. hoxbarana Bradfield	X	--------	? sp. indet	X	--------
? sp., aff. J. papei Scott	X	--------	Sansabella? sp	X	X
? sp. 1	X	--------	New genus aff. Gutschickia	X	
			Ostracoda, gen. indet		X
Malacostraca (?)					
Prawn (?), gen. undet	--------	X			
Blastoidea (?)					
Blastoid (?) plate, gen. indet	X	--------			

See footnotes at end of table.

TABLE 2.—*Fossils of the Belden and Paradox Formations of the Glenwood Springs, Colorado, area*—Continued

	Belden Formation	Paradox Formation		Belden Formation	Paradox Formation
Crinoidea					
Amphicrinus? sp. indet.	×	--------	*Parulocrinus* cf. *P. beedei* Moore and Plummer.		×
Erisocrinus typus Meek and Worthen		×	Crinoid columnals.	×	×
Hydriocrinus? sp. indet.	×	--------	Crinoid plates and spines.	×	
Mooreocrinus? n. sp.	×	--------	Crinoid (?) spines.	×	×
? sp., basal plate.	×	--------			
Echinoidea					
Echinocrinus cf. *E. dininnii* (White).	×	--------	Echinoid lantern plate (?).	×	--------
halliana (Geinitz).	×	×	Echinoid plates, gen. undet.	×	×
sp. A, aff. *E. halliana* (Geinitz).	×	--------	Echinoid spines, gen. undet.	×	×
sp. B.	×	--------	Echinoid (?) spines.	×	×
sp.	×	×			
Vertebrata					
Elonichthys sp.	--------	×	Spine fragments, gen. indet.	×	--------
Megalichthys sp.	×	--------	Tooth fragments, gen. indet.	×	×
Peripristis semicircularis Newberry and Worthen.	×	--------			
Petalodus ohioensis Safford.	--------	×	Total number of species.	258	103

[1] Cited by M. L. Thompson (1945, p. 23, fig. 2) from the Belden Formation on Sweetwater Creek. (See also op. cit., p. 42–43, 44–49, pls. 1,5.)
[2] = *Echinaria* sp. (Muir-Wood and Cooper, 1960).
[3] = *Desmoinesia* cf. *D. muricatina* (Dunbar and Condra) (Muir-Wood and Cooper, 1960).
[4] = *Hystriculina wabashensis* (Norwood and Pratten) (Muir-Wood and Cooper, 1960).
[5] Cited by G. H. Girty (1903, p. 233) from the Pennsylvanian rocks at Glenwood Springs.
[6] Cited by G. H. Girty (1903, p. 456) from the Pennsylvanian rocks at Glenwood Springs.
[7] Cited by G. H. Girty (1903, p. 458) from the Pennsylvanian rocks at Glenwood Springs.
[8] Described at length but not figured by G. H. Girty (1903, p. 324–327).

Algae, foraminifers, bryozoans, brachiopods, pelecypods, gastropods, annelids, and ostracodes are all locally abundant in the Belden strata. Less common are corals, crinoids, and echinoids. Scaphopods, cephalopods, trilobites, and vertebrates are rare.

An abundance of *Millerella* and *Paramillerella*, together with an absence of more advanced fusulinids, suggests a Morrow age. L. G. Henbest (written communication, May 1960), however, states,

Few if any would agree that the absence of a kind of fossil is reliable evidence on age. Furthermore the millerellids reached their acme in late Morrow or, more likely, Atoka or Bend time.

Profusulinella and *Fusulinella* characterize Atoka age, while *Wedekindellina* and *Fusulina* characterize the early Des Moines. *Millerella* and *Paramillerella* are abundant in the Belden Formation of the Glenwood Springs area. Of more advanced fusulinids, L. G. Henbest (written communication, June 30, 1952) found but a single specimen of a form he identified questionably as *Profusulinella*? sp. "or possibly the juvenarium of *Fusulinella*." The bed yielding this sample at a locality on Boiler Creek is assigned to the Belden. If the form is *Profusulinella*, then the age of the rock is probably earliest Atoka. Many samples from the Belden Formation at several localities yielded a variety of genera of nonfusulinid foraminifers. The absence of *Eoschubertella*, *Fusulinella*, *Pseudostaffella*, and *Staffella* is noteworthy if much of the Belden is Atoka age.

In her detailed report on the bryozoans, Helen Duncan (written communication, Mar. 27, 1951) concluded:

> The Belden bryozoan assemblage has a good many points of similarity with the Morrow faunule. Both faunas contain forms of *Fenestella*, *Cystodictya*, rhomboporoids, *Dictyocladia*, *Prismopora*, and the laminar and incrusting fistuliporoids and stenoporoids that are superficially similar if not specifically identical. On the other hand, the Belden contains a few bryozoans such as *Cheilotrypa*, *Meekopora*, and *Ramiporalia* that have not been reported from the Morrow. * * * The bryozoan assemblage found in the Belden contains some Des Moines elements, but it is not a faunule that I would consider exactly typical of the Pennsylvanian in Colorado—Hermosa, Maroon, Robinson, etc. In some ways it looks more like a Mississippian assemblage.

Among the brachiopods of the Belden, the following are notably abundant:

> *Antiquatonia coloradoensis*
> *Chonetinella flemingi*
>   *flemingi crassiradiata*
> *Composita ovata*
> *Derbyia crassa*
> *Marginifera ingrata*
> *Orthotichia schuchertensis*
> *Paeckelmannia* n. sp. aff. *P. derelicta*
> *Punctospirifer kentuckiensis*
> *Spirifer occidentalis*

The presence in the Belden of such forms as *Chonetinella flemingi*, *C. flemingi crassiradiata*, *Composita ozarkana*, *Horridonia bullata*, *Hustedia miseri*, *Lissochonetes* cf. *L. geinitzianus*, *Marginifera ingrata*, *M.* cf. *M. muricatina*,[2] *Orthotichia schuchertensis*, *Spirifer occidentalis*, and *S.* cf. *S. opimus* suggests a Morrow age. As M. K. Elias (1957, p. 516) has pointed out, *Spirifer occidentalis* generally appears earlier than *S. rockymontanus*. The former occurs in the Belden, while the latter occurs in the Paradox. A productoid quite similar to *Marginifera ingrata* of the Belden occurs in the Redoak Hollow Formation (Late Mississippian) of Oklahoma, along with *Spirifer opimus* and *Punctospirifer kentuckiensis*, according to Elias (1957, p. 499).

As noted above, the lower part of the Belden Formation is assigned to the Morrow and the upper part to the Atoka age.

---

[2] = *Desmoinesia* cf. *D. muricatina* (Muir-Wood and Cooper, 1960).

PARADOX FORMATION

The Paradox, which the authors consider a formation in this area, for the most part consists of thick beds of gypsum interbedded with units of fissile black shale. The thicker gypsum beds range from 65 to 160 feet. A well drilled in 1960 in the NW¼SE¼ sec. 12, T. 7 S., R. 89 W., 3 miles south of the map area, penetrated interbedded salt and gypsiferous siltstone 480 feet thick in the Paradox Formation at a depth of 2,125 feet, underlain by solid salt 455 feet thick to the bottom of the hole. (Data from well log prepared from drill samples by American Stratigraphic Co., Denver, Colo.) A sequence of reddish gypsiferous siltstone and shale, about 500 feet thick, is present near the middle of the formation. North of Dotsero the uppermost 100 feet of the formation grades upward through interbedded brown and yellow sandy shale and shaly sandstone, and gray shale into the overlying red beds of the Maroon Formation. The thickness of the formation is 1,553 feet at Blowout hill on the east side of the Colorado River, across the river from the mouth of Deep Creek, which is the only place where a section of the formation could be measured. The following section was measured opposite the mouth of Deep Creek about 1¼ miles north of the bridge on U.S. Highway 6 over the Colorado River. Elsewhere the beds of the formation have been so contorted by flowage of the gpysum that the boundaries of the beds are not readily determinable.

Stratigraphic section of Paradox Formation on west slope of Blowout hill, on east bank of Colorado River, 2 miles north of Dotsero, in secs. 28 and 29, T. 4 S., R. 86 W.

Maroon Formation (Pennsylvanian and Permian):
    Red beds—sandstone, siltstone, shale, and some beds of conglomerate— form the top 900 ft of the hill. Transition beds, dominantly reddish but containing a considerable number of gray beds—sandstone, siltstone, shale, and some conglomerate—725 ft thick underlie the 900-ft sequence of red beds and constitute the lowermost part of the Maroon Formation.

Paradox Formation (Middle Pennsylvanian):

		Feet
20.	Shale, calcareous, black, and interbedded brown and gray calcareous sandstone and mudstone	55
19.	Siltstone, pink, and interbedded gray gypsum	45
18.	Gypsum, massive, white; dark gray where weathered	65
17.	Gypsum, gray and dark-gray, and interbedded gray-brown shale	81
16.	Gypsum, white; dark gray where weathered; interbedded minor amount of black shale	150
15.	Siltstone, shaly, and interbedded silty shale; micaceous, calcareous, greenish-gray; weathers light reddish brown; very calcareous hard mudstone at the base of the unit	90

    (Unit 15 through unit 12, whose total thickness is 513 ft, forms a reddish sequence in the middle of the Paradox Formation.)

Paradox Formation (Middle Pennsylvanian)—Continued

		Feet
14.	Siltstone, shaly, and interbedded silty shale; micaceous and calcareous; includes at the base a 10-ft unit of reddish-brown fine-grained micaceous and calcareous sandstone	100
13.	Shale, silty, micaceous, calcareous, greenish-gray; weathers dull reddish brown; includes 3–4 beds, each 6 in. to 1½ ft thick, of dense, hard, very dark gray limestone, which are widely distributed in the unit	280
12.	Siltstone, calcareous, micaceous, dull, grayish-brown; weathers slabby; includes a 1-ft bed of very limy dark-gray dense mudstone near the middle	43

Base of 513-ft reddish zone.

11.	Shale, dark-gray, gypsiferous; includes a few thin beds of hard, noncalcareous shale, one 2-ft bed of gypsum, and a few 3-in. beds of fine-grained micaceous tan sandstone	46
10.	Siltstone and interbedded shale, tan, red, and gray; forms ledges locally and slopes elsewhere	70
9.	Siltstone, tan, interbedded with black shale and black mudstone; in part calcareous; lower half contains more drab-brown shale, in part calcareous, than upper half	86
8.	Siltstone, calcareous, and a little sandstone; weathers yellow; forms a ledge	24
7.	Mudstone, shale, and siltstone, gypsiferous, noncalcareous, dark-gray	25
6.	Gypsum, massive, white, weathers dull gray; seems to be all gypsum, but surface is formed by hard gypsum crust which may conceal thin beds of shale	75
5.	Shale, gypsiferous, dark-gray and black	46
4.	Similar to unit 6	160
3.	Shale, black, and interbedded tan sandstone	9
2.	Conglomerate, steeply crossbedded, mainly poorly sorted, coarse and very coarse quartz grains; includes pebbles and cobbles of granite as much as 2¾ in. in diameter; micaceous, light tan; forms a prominent ledge capping a long shale slope formed mainly by the upper part of the underlying Belden Formation	65
1.	Gypsum and interbedded black shale (basal unit of the Paradox Formation)	38

Total thickness of the Paradox Formation_____ 1,553

Belden Formation (Pennsylvanian):
Shale, black, and interbedded gray conglomerate.

Two units of fossiliferous limestone, 50 to 75 feet apart stratigraphically, are present in the westernmost part of the map area and in the area adjacent on the west. Each limestone unit is 40 to 50 feet thick and the upper one is about 400 feet below the top of the formation. These beds are particularly well exposed on Main Elk Creek, 4 miles west of the map area, and one is exposed in a vertical attitude on Canyon Creek, within the map area.

A total of 42 lots of fossils were collected from the Paradox Formation and the fauna numbers 103 species. See table 2 for a list of these fossils.   Common to the Belden and Paradox are 58 forms; most of these are generic determinations or tentative or queried specific determinations, and only 14 definitely determined species are common to the two faunas, as follows:

> Antiquatonia coloradoensis
> Cleiothyridina orbicularis
> Composita ovata
> Derbyia crassa
> Hustedia mormoni
> Linoproductus prattenianus
> Orbiculoidea missouriensis
> Petrocrania modesta
> Punctospirifer kentuckiensis
> Bakewellia parva
> Myalina (Myalinella) cuneiformis
> Spirorbis sp. A
> Sevillia trinucleata
> Echinocrinus halliana

A striking contrast between the Belden and Paradox assemblages is the relative abundance in the Belden and scarcity in the Paradox of foraminifers, gastropods, and ostracodes.

Although locally abundant, the few kinds of calcareous algae afford no evidence of age.   Despite extensive search, few foraminifers were found in the Paradox Formation; these include undetermined calcitornellids and tolypamminids.   If, as believed, the Paradox is of Atoka and Des Moines age, the absence of such fusulinid genera as *Fusulinella*, *Wedekindellina*, and *Fusulina* is surprising.   The few corals found in the Belden are not diagnostic, but the Paradox contains a number of tentatively identified lophophyllidids.   Of these, one related form occurs in the Wapanucka Limestone (Morrow age) of Oklahoma; one occurs in the lower part of the Pottsville Formation of Ohio; two occur in rocks of Des Moines age (Wewoka Formation of Oklahoma and Strawn Group of Texas); and one occurs in the Kansas City Group (Missouri age).

Of the bryozoans, Helen Duncan (written communication, Mar. 27, 1951) remarks:

The poorly preserved bryozoan faunule is composed of forms commonly found in the Lower and Middle Pennsylvanian (Des Moines and older) rocks of Colorado. The species that occur are not conspicuously different from those found in the Belden, but certain genera that are characteristic of the Belden and of the Morrow and Pottsville faunas are scarce or absent.

Among the brachiopods, the following are notably abundant in the Paradox:

> Antiquatonia coloradoensis
> Composita ovata
> Derbyia crassa

Hustedia mormoni
Marginifera wabashensis [3]
Mesolobus mesolobus decipiens
Phricodothyris perplexa
Punctospirifer kentuckiensis
Spirifer rockymontanus
Wellerella tetrahedra

Of these, *Marginifera wabashensis,* [4] *Mesolobus mesolobus decipiens,* *Phricodothyris perplexa,* *Spirifer rockymontanus,* and *Wellerella tetrahedra,* together with *Neospirifer dunbari,* are restricted to the Paradox and suggest a Des Moines age, at least in part. *Mesolobus mesolobus decipiens* first appears in the lower part of the Des Moines Series of the Midcontinent and it does not range above the top of the series. *Phricodothyris perplexa,* extremely common in the Paradox, is also common in the Robinson Limestone Member of the Minturn Formation of the Kokomo area. This species is locally common in some of the Des Moines faunas of the Midcontinent region.

There is a notable paucity of gastropods and ostracodes in the Paradox. Of the few determinable crinoid remains, those in the Belden suggest a Morrow age and those in the Paradox suggest Des Moines to Missouri age, according to Bowsher (written communication, Jan. 22, 1951). Only a few fish teeth were found and these are not diagnostic.

The authors conclude that the Paradox fauna, collected chiefly from the uppermost 400 feet of the formation, is of early Des Moines age. Inasmuch as the fauna a few hundred feet below the Paradox Formation suggests an Atoka age, it is reasonable to conclude that the Paradox Formation in the Glenwood Springs area is of Atoka and Des Moines age. It has been noted by L. G. Henbest (written communication, May 1960) that

However correct the correlation [of the Paradox Formation] at the Glenwood Springs area with the type Paradox may be, the ages may actually differ from place to place. The absence of the fusulinids listed may be a result of (a) different age, or (b) unfavorable environment or some accident such as inaccessibility, plague, etc. It is not unusual to find fusulinid fossils missing from seemingly good lithologies of the right age.

In the Crested Butte quadrangle, to the southeast of Glenwood Springs, *Prismopora* sp., *Chonetinella flemingi, Lissochonetes* cf. *L. geinitzianus, Spirifer rockymontanus,* and *Spirorbis* sp. are found in the Belden Formation, according to Langenheim (1952, p. 566), whereas *Derbyia crassa, Hustedia mormoni, Linoproductus prattenianus, Mesolobus mesolobus, Neospirifer, Phricodothyris perplexa, Punctospirifer kentuckiensis,* and *Spirifer occidentalis* first appear in the

---

[3] = *Hystriculina wabashensis* (Muir-Wood and Cooper, 1960).
[4] = *Hystriculina wabashensis* (Muir-Wood and Cooper, 1960).

overlying Gothic Formation of Langenheim (1952) (supposedly Des Moines in age).

In the Minturn area, east of Glenwood Springs, the following forms have been found between the Robinson Limestone Member and the top of the Minturn Formation: *Antiquatonia coloradoensis, Chonetinella* cf. *C. flemingi, Hustedia* cf. *H. mormoni, Linoproductus prattenianus, Mesolobus mesolobus, Neospirifer, Spirifer* cf. *S. opimus,* and *S. rockymontanus* (Brill, 1942, p. 1388).

From his newly named Deer Creek Formation of the Sangre de Cristo Mountains of southern Colorado, Bolyard (1959, p. 1910) cited the following: *Lissochonetes geinitzianus, Neospirifer, Spirifer* cf. *S. occidentalis, S.* cf. *S. opimus,* and *S. rockymontanus.* From the overlying Madera Formation, Bolyard (1959, p. 1916–1917) cited *Marginifera wabashensis*[5] and *Phricodothyris perplexa.*

In the Nacimiento Mountains of north-central New Mexico, *Spirifer occidentalis* appears in Northrop's faunal zone A (Morrow in age), whereas such forms as *Prismopora, Antiquatonia coloradoensis, Derbyia crassa, Linoproductus prattenianus,* and *Spirifer rockymontanus* do not appear until faunal zone B (Atoka and Des Moines). Such species as *Hustedia mormoni, Neospirifer, Phricodothyris perplexa,* and *Punctospirifer kentuckiensis* do not appear until faunal zone C or higher (Wood and Northrop, 1946).

According to Gehrig (1958, p. 8–9), in southern New Mexico *Spirifer occidentalis* appears in the Derry Series of Thompson (1942) and continues upward into Des Moines time. *Phricodothyris perplexa* first appears in the top beds of the Derry Series but is more abundant in the Des Moines. Again, *Mesolobus mesolobus, Neospirifer,* and *Spirifer rockymontanus* do not appear in this area until Des Moines time.

As noted above, the authors conclude that the lower several hundred feet of the Belden Formation in the Glenwood Springs area is of Morrow age and that the upper part is of Atoka age. The Paradox Formation in the Glenwood Springs area is probably of late Atoka and Des Moines age.

PENNSYLVANIAN AND PERMIAN

MAROON FORMATION

As used herein the Maroon Formation includes a thick sequence of red beds between the gypsum-bearing Paradox Formation below and the top of a dull-maroon shale, 50 to 100 feet thick, that overlies a thin dolomite which contains a Phosphoria fauna, and underlies the orange-red rocks of the Chinle Formation. The formation is about 3,350 feet thick near Glenwood Springs and on Main Elk Creek, 4 miles

---

[5] = *Hystriculina wabashensis* (Muir-Wood and Cooper, 1960).

west of the map area. The data available suggest that the formation has a comparable thickness on the west and northwest flanks of the White River uplift, but that the thickness is much less on the east and southeast flanks. The thickness was not measured here, but the formation is less than 1,000 feet thick on Eagle River, 6 miles northeast of Eagle, which is 13 miles east of the map area.

The Maroon Formation consists of predominantly red even-bedded shale, siltstone, sandstone, and conglomerate, and a few thin beds of dark-gray dense limestone. Many beds are arkosic and most are micaceous. Beds of siltstone or fine- to coarse-grained sandstone and beds of conglomerate containing pebbles and cobbles as large as 2 to 3 inches, and rarely 5 inches in diameter, alternate with beds of silty shale. Many beds of siltstone and fine-grained sandstone contain lenses of coarse sand and gravel. Some silty beds contain casts of mud cracks. In general, the finer grained beds are darker red than the coarser grained beds. Some beds of coarse sandstone are light gray. Many beds of coarse conglomerate are purplish red.

The exact age of much of the Maroon Formation in unknown. Fossils of early Des Moines age are present in the western part of the area in limestone beds 400 feet below the base of the formation and a Permian (Phosphoria) fauna is present 50 to 100 feet below the top. Accordingly, the Maroon Formation is of both Pennsylvanian and Permian age.

TONGUE OF WEBER SANDSTONE

A gray sandstone, about 400 feet thick, that lies 100 to 270 feet below the top of the Maroon Formation and crops out in the southwestern part of the area and for many miles west and south of the map area (Donnell, 1954), is believed to be a tongue of the Weber Sandstone of northwestern Colorado (Thomas, McCann, and Raman, 1945). The sandstone alternates from a solid gray unit to interbedded red and gray. The grain size is variable; much of it is fine to medium. Some beds contain poorly sorted quartz grains ranging from fine to coarse, and a few thin beds contain pebbles whose maximum diameters are 1¼ inches. The rock is slightly micaceous. The sandstone is impregnated with a dull-black organic substance at all outcrops, which imparts a very dark gray color to freshly fractured surfaces. Analysis by the Geological Survey showed the substance to be a residue of petroleum.

The age of the Weber Sandstone, whose thickness is more than 1,000 feet in the Uinta Mountains, northeastern Utah, and northwestern Colorado (100 miles northwest of the map area), is now classed as Pennsylvanian and Permian (L. G. Henbest, written communication, Sept. 29, 1961). Southeastward from the Uinta Mountains the Weber interfingers with arkosic red beds of the Maroon Formation. It appears probable that it is the upper part of the Weber that persists southeastward to form the tongue of the Weber Sandstone (Bissell and Childs, 1958) in the map area. Its stratigraphic position in the upper part of the Maroon, only a short interval below the South Canyon Creek Member, which contains a Permian fauna, suggests that it may be equivalent to some part of the Permian portion of the Weber Sandstone.

SOUTH CANYON CREEK MEMBER

A fossiliferous gray dolomite and limestone unit, lying about 50 to 100 feet below the top of the Maroon Formation, was named the South Canyon Creek Dolomite Member of the Maroon Formation (Bass and Northrop, 1950, p. 1541–1542) from its clean exposures on South Canyon Creek, 4½ miles west of Glenwood Springs. It is now called the South Canyon Creek Member (Hallgarth, 1959). It was identified and fossil collections were made from it at many places extending from an outcrop a few hundred yards east of South Canyon Creek northwestward for 22 miles to Middle Rifle Creek, which is 14 miles west of the map area.

The fauna is a typical facies fauna, exclusively molluscan and composed of 19 pelecypods, 2 scaphopods, and 3 gastropods, as follows:

Pelecypoda:
    Allorisma cf. *A. rothi* Newell
      ? sp.
    Aviculopecten sp.
    Culunana? sp.
    Deltopecten? sp.
    Edmondia? sp.
    Lima? sp.
    Myalina (*Myalina*) cf. *M.* (*M.*) *wyomingensis thomasi* Newell
      (*Myalinella*) cf. *M.* (*M.*) *acutirostris* Newell and Burma
      sp.
    Nucula? sp.
    Pleurophorus cf. *P. albequus* Beede
      cf. *P. mexicanus* Girty
      sp.
    Schizodus cf. *S. ferrieri* Girty
      sp. (abundant)
    Solenomya sp.
      ? sp.
    Streblochondria? sp.
Scaphopoda:
    Plagioglypta canna White
      monolineata Branson
Gastropoda:
    Bellerophon sp.
    Euomphalus sp.
    Naticopsis? sp.

Northrop (Bass and Northrop, 1950, p. 1548) has compared this fauna with a number of Permian faunas from Idaho, Wyoming, Utah, Arizona, New Mexico, Texas, and Oklahoma. The fauna seems to be of Phosphoria and Kaibab age (Bass and Northrop, 1950, p. 1550–1551).

The thickness of the South Canyon Creek Member ranges from 18 inches to 6 feet 4 inches. The lower part of the member consists of fine- to medium-grained dolomite and dolomitic limestone. It commonly contains a few veinlets and irregular nodules of milk-white or light-gray chert, and vugs and geodes, most of which are less than 1½ inches in diameter, and many of which contain jet-black brittle lustrous asphalt. The upper part of the member consists of thinly laminated limestone which has a wavy structure and alternating light and dark bands which suggest an algal origin for the rock. Where recrystallization has occurred, laminae of finely crystalline calcite are bounded by very thin seams of asphalt. Seams and nodules of chert are present locally.

Stratigraphic section of the Maroon Formation on west side of Main Elk Creek, about 4 miles west of the map area, in secs. 2, 3, and 10, T. 5 S., R. 91 W., and sec. 35, T. 4 S., R. 91 W.

Chinle Formation (Upper Triassic):
  Interbedded orange-red shale, siltstone, and limestone-pebble
    conglomerate, and some sandstone.

Maroon Formation (Pennsylvanian and Permian):	Ft	In
87. Shale, dull dark-maroon	50	0
South Canyon Creek Member (units 86 to 83):		
86. Limestone, gray, fossiliferous, crenulated bedding (probably algal); contains geodes with asphalt particles, and a little milk-white chert	1	4
85. Dolomite, light-gray, fine-grained, massive; forms a single ledge	2	8
84. Dolomite, light-gray, fine-grained; weathers with a chippy and rounded edge	1	5
83. Dolomite, light-gray, and fine- to medium-grained dolomite and dolomitic limestone, in part silty	1	0
Thickness of South Canyon Creek Member	6	5
82. Siltstone and shale, calcareous, dull, brownish-red, in part mottled-gray	48	0
Tongue of Weber Sandstone (Permian) (units 81 to 72):		
81. Sandstone, very fine grained to silty, micaceous, calcareous, thin-bedded, light-gray	15	0
80. Sandstone, in part shaly, in part concealed; fine- to medium-grained, light-gray	20	0
79. Sandstone, fine-, medium-, and coarse-grained, arkosic, calcareous, gray	15	0

Maroon Formation—Continued
  Tongue of Weber Sandstone (Permian)—Continued

	Ft	In
78. Sandstone, micaceous, arkosic, calcareous, poorly sorted, fine-grained, thin-bedded, gray; contains dead-oil stain	33	0
77. Covered; probably mostly sandstone and siltstone	65	0
76. Sandstone, fine-grained, arkosic, micaceous, calcareous, thin-bedded, gray; weathers yellow	25	0
75. Mostly covered; some siltstone and shaly sandstone exposed	15	0
74. Sandstone, coarse-grained, poorly sorted, crossbedded, arkosic, micaceous, calcareous; contains dead-oil stain	10	0
73. Siltstone, micaceous, red, and interbedded, thin-bedded, coarse, arkosic, micaceous, and calcareous sandstone; includes a 3-ft bed of light-gray sandstone in the middle	49	0
72. Sandstone, fine- to coarse-grained, micaceous, calcareous; color ranges from light gray to pink to brown	43	0
Thickness of tongue of Weber Sandstone	290	0
71. Sandstone, fine- to coarse-grained, micaceous, calcareous, light-gray to pink and red- and gray-mottled	9	0
70. Conglomerate, crossbedded; pebbles of quartz commonly less than ½ in. in diameter but as much as 1 in.; calcareous, dull, maroon	9	0
69. Sandstone, medium-grained, micaceous, calcareous, mostly light-red, some light-gray	42	0
68. Conglomerate, arkosic, crossbedded; quartz pebbles as much as 1 in. in diameter; in part covered	70	0
67. Sandstone, fine-grained, micaceous, calcareous, red; in thin layers ¼–1 in. thick	80	0
66. Conglomerate in upper part, with quartz and feldspar pebbles as much as ½ in. in diameter, light-tan to gray; grades downward to interbedded fine- and medium-grained, micaceous, calcareous, red sandstone and siltstone	65	0
65. Conglomerate and interbedded sandstone, with pebbles as much as 2¼ in. in diameter; arkosic, micaceous, calcareous; red	82	0
64. Sandstone, coarse-grained, interbedded with medium- and fine-grained sandstone and siltstone; arkosic, micaceous, calcareous, red	80	0
63. Sandstone, medium- to coarse-grained, and interbedded siltstone; arkosic, micaceous, calcareous, brownish-red; forms a cliff	30	0
62. Siltstone and interbedded medium- to coarse-grained sandstone; micaceous, calcareous, red	47	0
61. Sandstone, fine- to medium-grained, and some siltstone; micaceous, red	12	0

Maroon Formation—Continued

	Ft	In
60. Sandstone, calcareous, red and lavender; much is coarse grained and arkosic; some is medium to coarse grained	116	0
59. Conglomerate, with pebbles as much as 1 in. in diameter, arkosic, micaceous, calcareous, dull-red with pink and light-gray streaks	50	0
58. Siltstone, in part laminated, slightly calcareous, micaceous, bright-red	57	0
57. Sandstone and conglomerate; top 25 ft and lower 40 ft are conglomerate; rest is mainly sandstone, with a little siltstone; pebbles in conglomerate as much as 1 in. in diameter; arkosic, micaceous, calcareous, red and pink	127	0
56. Concealed	10	0
55. Sandstone, coarse-grained, arkosic, calcareous, gray	12	0
54. Mostly concealed; probably red, fine-grained calcareous sandstone	30	0
53. Sandstone and conglomerate; pebbles as much as 1 in. in diameter; arkosic, micaceous, calcareous, red	20	0
52. Sandstone, fine- to medium-grained, poorly sorted and interbedded, micaceous, arkosic, calcareous, red	88	0
51. Conglomerate, crossbedded, arkosic, micaceous, calcareous; pebbles commonly less than ½ in., but as much as 2 in. in diameter	51	0
50. Siltstone, thin-bedded, and some beds of fine- to medium-grained sandstone; micaceous, calcareous, red	15	0
49. Sandstone, fine- to coarse-grained, subangular, poorly sorted, arkosic, micaceous, calcareous, reddish-gray	43	0
48. Sandstone and interbedded conglomerate; in descending order, 3 ft of conglomerate, 20 ft of sandstone, 5 ft of conglomerate, and 25 ft of conglomerate; the sandstone locally has contorted bedding; the top conglomerate contains limestone pebbles 3 in. in diameter; otherwise the sandstone and conglomerate are similar to units 53-51; the basal sandstone has thin-bedded units of red fine-grained sandstone alternating with light-gray beds about ½ in. thick of medium-grained poorly sorted sandstone	53	0
47. Sandstone, very coarse grained, and conglomerate; crossbedded, arkosic, micaceous, calcareous, red; containes a few pebbles 2 in. long	90	0
46. Siltstone, thin-bedded, micaceous, calcareous, red; contains a few 1-in. beds of coarse-grained red sandstone	40	0
45. Sandstone and conglomerate, massive, arkosic, micaceous, calcareous, subangular grains, red	60	0
44. Sandstone, fine- to very coarse-grained, arkosic, micaceous, calcareous, gray	50	0

Maroon Formation—Continued

	Ft	In
43. Sandstone similar to unit 44, interbedded with micaceous, calcareous, red siltstone_____	65	0
42. Sandstone, massive, medium- to coarse-grained, micaceous, calcareous, red; forms most prominent ledge about 75 ft above the road_____	33	0
41. Siltstone and interbedded sandstone; micaceous, calcareous, red; lower part mostly siltstone; unit partly concealed_____	50	0
40. Sandstone, interbedded with shaly sandstone and siltstone; micaceous, calcareous, red with gray patches; some beds are gray; thick unit of siltstone in upper part_____	74	0
39. About 50 percent concealed; interbedded sandstone siltstone, silty shale, and sandy red shale_____	140	0
38. Much concealed; seems to be interbedded coarse-grained sandstone, fine-grained sandstone, shale, silty shale, and sandy red shale_____	85	0
37. Much concealed, but in general a shaly sequence; some beds of red and gray sandstone and siltstone exposed, as well as 3 beds of gray limestone; the lowest bed of limestone is 2½ ft thick, is crystalline, and is 18 ft above the base of the unit; the upper 2 beds of limestone are each 1 ft thick, dense and gray and are near the middle of the unit; the interval between them is 20 ft; the lowest 10 ft of the unit is coarse-grained micaceous calcareous red sandstone_____	45	0
36. Sandstone, and minor amounts of shaly sandstone; fine- to medium-grained, micaceous, slightly calcareous, red to dull brownish-red_____	25	0
35. Sandstone interbedded with shaly sandstone and silty shale; much mica, calcareous, red (in general a shaly unit)_____	66	0
34. Siltstone and interbedded sandstone; calcareous and gray; includes several thin beds of gray shaly limestone_____	9	0
33. Shale, gray, and some thin beds of gray shaly limestone_____	18	0
32. Sandstone, micaceous, medium- to coarse-grained, calcareous, gray_____	8	0
31. Shale, limy, gray, interbedded with thin beds of gray shaly limestone and some sandstone; basal 10 ft contains a few beds 6–8 in. thick of dense, gray limestone and gray shaly limestone; some limestone beds have tiny ripple marks on surface_____	37	0
30. Limestone, dense, gray; weathers into a ledge with a jagged surface, and blocks with wavy surface_____	15	0
(Units 34 through 30 form a gray sequence within the red beds.)		
29. Concealed_____	4	0

Maroon Formation—Continued

	Ft	In
28. Conglomerate, crossbedded, gray; fragments of underlying siltstone in basal few inches	8	0
27. Sandstone and interbedded shaly sandstone; red	30	0
26. Conglomerate, coarse and very coarse grains of quartz, feldspar, mica; slightly calcareous, gray	15	0
25. Sandstone, coarse-grained, and shaly sandstone; red	12	0
24. Conglomerate, micaceous, arkosic, reddish-brown	13	0
23. Sandstone, coarse- to medium-grained, crossbedded, micaceous, calcareous, red, and interbedded finer grained evenly bedded red sandstone; a few beds of the coarse-grained sandstone are gray	115	0
22. Sandstone, coarse, arkosic, micaceous, slightly calcareous, reddish-brown	20	0
21. Sandstone, micaceous, arkosic, coarse-grained, red, in massive beds, and interbedded red siltstone in beds about 8 ft thick	75	0
20. Conglomerate, chiefly coarse and very coarse subangular quartz grains and some pebbles; micaceous	10	0
19. Siltstone and interbedded silty shale; orange-red	9	0
18. Conglomerate similar to unit 20	8	0
17. Siltstone and interbedded silty shale; micaceous, red	28	0
16. Sandstone, slabby-bedded, micaceous, calcareous, red; includes in lower part a few thin beds of conglomerate composed chiefly of coarse sand	50	0
15. Conglomerate, crossbedded, in thick beds; contains many coarse and very coarse subangular quartz grains; arkosic, micaceous, calcareous	35	0
14. Siltstone, very shaly, and interbedded silty and sandy shale; micaceous, finely laminated, calcareous, red; upper boundary is a sharp contact with overlying conglomerate	15	0
13. Sandstone, fine- to medium-grained, micaceous, slabby-bedded, red	14	0
12. Shale, red, with a few 1-in. beds of dense gray limestone	10	0
11. Sandstone, coarse-grained, subangular, micaceous, calcareous, light-gray	4	0
10. Sandstone, in slabby beds, fine-grained, micaceous, dull, gray; in part shaly; 10 percent concealed	15	0
9. Limestone, dense, slabby, gray	1	8
8. Sandstone, coarse-grained, subangular, calcareous, micaceous, gray	38	0
7. Sandstone, medium-grained, arkosic, micaceous, calcareous, dull, maroon	20	0
6. Conglomerate, and coarse and very coarse grained sandstone; some pebbles as much as ½ in. in diameter; mostly subangular quartz grains, arkosic, micaceous, slightly calcareous; 20 percent concealed	56	0
5. Sandstone, coarse quartz grains, micaceous, arkosic; 25 percent concealed	52	0

Maroon Formation—Continued

		Ft	In
4.	Sandstone, silty, fine-grained, micaceous, calcareous, interbedded light-gray and red beds............	12	0
3.	Sandstone, coarse-grained micaceous, arkosic, calcareous, light-gray, and interbedded calcareous micaceous red siltstone.....................	40	0
2.	Sandstone, medium- to coarse-grained, and interbedded fine- to medium-grained sandstone, coarse-grained sandstone, and siltstone; in part laminated, micaceous, arkosic, calcareous..................	53	0
1.	One-half or more concealed; interbedded micaceous, calcareous siltstone and sandstone.............	90	0
		2,959	6
	Total thickness of Maroon Formation..........	3,354	—

Paradox Formation (Middle Pennsylvanian) (top unit only):

	Ft	In
Limestone, in 4 beds with shale partings, middle part cherty; dark- to light-gray; basal part is very argillaceous.........	7	0

UPPER TRIASSIC

CHINLE FORMATION

A sequence about 325 feet thick of interbedded orange-red shale, siltstone, and limestone-pebble conglomerate constitutes the Chinle Formation. Its contact with the underlying Maroon Formation is arbitrarily drawn at the contact of a dull maroon shale (below) with an overlying orange-red silty to sandy shale. The Shinarump is the lowest member of the Chinle Formation in parts of northwestern Colorado and Utah, but it was not observed here. The upper boundary of the Chinle Formation is clearly defined by its disconformable contact with the overlying Entrada Sandstone—a light-gray rock.

UPPER JURASSIC

ENTRADA SANDSTONE

The Entrada Sandstone, about 100 feet thick, consists mostly of very fine to fine well-sorted subangular grains of clear quartz bonded with a slightly calcareous cement to form a firm rock that is locally crossbedded. Parts of the formation contain scattered rounded sand grains of medium to coarse size. The Entrada Sandstone forms a prominent light-gray ledge at nearly all exposures and is particularly conspicuous because of its position above the red rocks of the Chinle Formation.

MORRISON FORMATION

The Morrison Formation is 480 to 600 feet thick. It consists of pale-green shale interbedded with maroon shale, light-gray sandstone, and a few beds of dark-gray limestone. The beds of sandstone are most numerous in the lower hundred feet of the formation. They consist largely of fine quartz grains and contain a greater proportion of red, green, and brown grains than the underlying Entrada Sandstone. Thin beds of limestone, some of which are earthy to nodular, are present at several horizons in the formation; the thickest sequence of limestone is 140 feet above the base of the formation.

The limestone is medium dark gray and dense and contains in abundance specimens of Charophyta, described by R. E. Peck (1957) as a new genus and new species, *Echinochara spinosa*. The holotype and numerous paratypes, collected by Bass on South Canyon Creek in sec. 2, T. 6 S., R. 90 W., are described and figured by Peck (1957, p. 21–24; pl. 1, figs. 4, 9, 11–14, 16–17, 19, 21; pl. 2, figs. 23, 25). The same species occurs in the Morrison Formation at Perry Park, Colo.; north of Fort Collins, Colo.; north of Medicine Bow, Wyo.; and north of Edgemont, S. Dak. The species is closely related to *Clavator pecki* Mädler from the Kimmeridgian of Germany and may be the same species (Peck, 1957, p. 24).

LOWER(?) AND UPPER CRETACEOUS

DAKOTA SANDSTONE

The Dakota Sandstone, about 150 feet thick, generally is made up of two thick units of sandstone separated by a unit of sandy shale and shale. At some places three sandstone units are present and at other places two sandstone units are merged into a single unit, but the middle part is generally more friable than the upper and lower parts. The lower part of the Dakota is a sandstone and chert conglomerate. The chert pebbles are light gray and dark gray; most are less than one-half inch in diameter and some attain a diameter of 1 inch. The pebbles are contained in a matrix of fine to coarse quartz sand. The upper unit of the formation consists of clean subangular to angular grains ranging in size from very fine to medium, but is well sorted in individual beds. The Dakota forms prominent hogbacks; these are particularly well developed on the southwestern and western flanks of the White River uplift.

UPPER CRETACEOUS

MANCOS SHALE

The Mancos Shale consists essentially of gray shale. As measured by two planetable sections, it is a little more than 5,000 feet thick. A section measured 1¼ miles northwest of New Castle, west of the map area, suggests a thickness of 5,400 feet and one measured on South Canyon Creek suggests a thickness of 5,000 feet. Little confidence can be placed in the precise thickness figures based on the planetable surveys, however, because the shale is poorly exposed and the outcrops of the lower part of the formation are separated by nearly a mile from those of the upper part. The Mancos is not separated into members in this area, although certain members could be mapped.

A sequence of rock 50 feet thick, composed of interbedded siliceous shale and gray shale containing fish scales, is exposed about 3 feet above the base of the Mancos Shale, 1¾ miles northwest of New Castle east of the Buford road, which is 2½ miles west of the map area, and on South Canyon Creek, 5 miles west of Glenwood Springs. These beds were tentatively identified by W. A. Cobban as equivalent to the Mowry Shale of northwestern Colorado, which has been assigned an Early Cretaceous age (Cobban and Reeside, 1951). A unit 120 feet thick, composed of dark-gray shale, overlies the siliceous shale unit. Overlying the shale unit is a ridge-forming unit 50 feet thick, composed of slabby-bedded brown fossiliferous limy sandstone and sandy shale, which is probably equivalent to a part of the Frontier Formation. A dark-gray shale unit 70 feet thick overlies the Frontier equivalent and in turn is overlain by a sequence—several hundred feet thick, of light-gray limy shale interbedded with thin beds of light-gray chalky limestone—that is equivalent to the Niobrara Formation. Although the base of the Niobrara equivalent is apparent, the top boundary is not mappable because the upper part grades into the overlying somewhat darker gray shale. The Niobrara equivalent forms a broad light-gray band on the slopes which is readily visible on aerial photographs.

Although the Cretaceous rocks are locally fossiliferous, little systematic collecting was attempted. Several small collections were made by Northrop in 1947 from the Mancos Shale where it is exposed along the west wall of the canyon of South Canyon Creek, in T. 6 S., R. 90 W., west of Glenwood Springs. According to W. A. Cobban (written communication, Nov. 9, 1949), three collections (USGS Mesozoic localities 21651, 21652, 21653) contain the following:

> *Inoceramus dimidius* White
> *Ostrea lugubris* Conrad
> *Prionocyclus wyomingensis* Meek
>   cf. *P. macombi* Meek
> *Scaphites* sp.
> Fish bones and scales
> Tracks and burrows, indet.

These fossils suggest a middle Carlile age.

From a zone about 10 feet below the base of the Niobrara equivalent (loc. 21654), the following were collected:

> *Inoceramus perplexus* Whitfield
> *Baculites* n. sp. (smooth)
> *Prionocyclus* sp.
> Fish scale

These fossils suggest a Carlile age.

From the Niobrara equivalent (loc. 21655), the following were obtained:

Inoceramus deformis Meek
Ostrea congesta Conrad

The age of this collection is early Niobrara.

Cretaceous fossils were collected from several localities in T. 5 S., R. 91 W., northwest of New Castle, west of the map area.

USGS locality 21656, Mancos Shale within 5 to 10 feet of the Dakota:
Fish scales
USGS locality 21657, Macos Shale along New Castle-Buford highway 0.7 mile northwest of junction with East Elk Creek road:
Inoceramus dimidius White
Scaphites sp.
Age: middle Carlile.
USGS locality 21658, Mancos Shale near last locality:
Inoceramus deformis Meek
Age: early Niobrara.

Another small collection (loc. 23219), from the Mancos Shale at the first hogback west of the New Castle-Buford highway near the road junction in sec. 16, T. 5 S., R. 91 W., contains numerous specimens of *Inoceramus dimidius* White, an index fossil to rocks of middle Carlile age (W. A. Cobban, written communication, Jan. 15, 1951).

On Aug. 7, 1951, Bass and Northrop observed abundant foraminifers in gray slabby shaly limestone of the Niobrara equivalent at quarries of an abandoned cement plant in sec. 30, T. 5 S., R. 90 W., north of New Castle.

MESAVERDE GROUP

The Mesaverde Group is about 5,300 feet thick, according to a planetable survey made on Coal Ridge in the southwestern part of the map area. It consists of interbedded tan sandstone, tan and gray sandy shale, shale, and coal. Many beds of sandstone are massive and form prominent ledges.

The coal beds are present in three groups: lower, middle, and upper. The Wheeler, *D*, and Allen beds are well-known coal beds in the lower group. The Wheeler bed lies 900 feet above the base of the Mesaverde, and the *D* and the Allen beds are 85 and 600 feet, respectively, above the Wheeler bed. Three coal beds of workable thickness, known as *C*, *B*, and *A* in ascending order, are present in the middle group at the place where the planetable section was measured. Beds *B* and *A* lie at intervals of 45 and 130 feet above the *C* bed, and the three are about 2,200 feet, stratigraphically, above the base of the Mesaverde. One coal bed in the upper, or Keystone coal group, lies 3,900 feet above the base of the Mesaverde in the section measured.   Other thin coals are present in the upper group but are concealed at the site of the measured section.   Some of these, however, are exposed on South Canyon Creek.

PALEOCENE AND EOCENE

OHIO CREEK CONGLOMERATE, UNNAMED UNIT, AND WASATCH FORMATION

The Wasatch Formation and related beds overlying the Mesaverde Group are poorly exposed in a broad belt in the southwesternmost part of the area and are better exposed a few miles west of the area.   This sequence of rocks includes beds of Paleocene and Eocene age.   It is about 5,000 feet thick and consists of interbedded varicolored clay and clay shale, lenticular sandstone, and conglomerate—likely all of fluviatile origin.   A cobblestone conglomerate, consisting largely of subrounded cobbles of metamorphic and igneous rocks, is present in the lower part.   It is underlain by 40 to 115 feet of beds having at the base a massive light-gray sandstone that contains pebbles of brown, red, and black chert.   This conglomeratic sandstone has been identified by J. R. Donnell as the Ohio Creek Conglomerate.   Plant and vertebrate remains of Paleocene age are present 200 feet above the Ohio Creek Conglomerate in the south part of the Piceance Creek basin, according to Donnell (oral communication).   As thus defined, a sequence of sandstone, shaly sandstone, and drab clay shale underlies the Wasatch and constitutes the Ohio Creek Conglomerate and an overlying unnamed unit of Paleocene age.   About 1¼ miles southwest of New Castle and 2,000 feet north of U.S. Highway 6, 3 miles west of the map area, the Ohio Creek Conglomerate is the uppermost unit in a sandstone sequence, 200 feet thick, which forms three prominent bare ledges that dip southwest; the upper and lower units are light gray and the middle unit is maroon.   The identification of the Ohio Creek Conglomerate and overlying beds of Paleocene age was not made while fieldwork was in progress and these beds were mapped with the Wasatch Formation.

MIOCENE(?)

WHITE SAND

A body of white loosely consolidated sand, which is at least 150 feet thick, underlies the main basalt flows on Dome Peak in T. 1 S., R. 86 W., near the northeast corner of the area, and probably is present elsewhere in the northeastern part of the map area.   Parts of it seem to be massively bedded, and parts are intricately crossbedded at angles ranging up to 30°.   Similar white sand is exposed near Twin Lakes in the Flat Tops in the western part of T. 1 S., R. 88 W.   It seems possible that the sand is equivalent to a part of the Browns Park Formation, which is of Miocene(?) age and is widespread in northwestern Colorado.

LATE TERTIARY OR PLEISTOCENE

CONGLOMERATE ON CANYON CREEK

A coarse cobblestone conglomerate whose exact age is unknown is present on Canyon Creek about 1½ miles above its junction with the Colorado River. The conglomerate, estimated to be about 200 feet thick on Canyon Creek, is steeply folded into a syncline, where it is exposed in a southeastward-trending belt about 4 miles long and less than half a mile wide.

The conglomerate is composed of stream-laid cobbles of red sandstone, red siltstone, gray limestone (some of which is oolitic), chert, quartzite, and igneous and metamorphic rocks. It lies on gypsum beds of the Paradox Formation and is overlain by a large lateral moraine, which, according to G. M. Richmond (written communication, 1951), is probably of the Wisconsin stage. Cobbles range from 2 to 11 inches in diameter. Cobbles derived from the Sawatch, Leadville, Belden, and Maroon Formations were tentatively identified. Rocks younger than the Maroon Formation were not found. These facts suggest that the conglomerate was deposited much later than the orogeny that produced the White River uplift and prior to the deposition of the moraine of Wisconsin age. Thus, it appears probable that the age of the conglomerate is late Tertiary or Pleistocene.

The attitude of the conglomerate in an elongate syncline was probably produced by subsidence into caverns in the underlying thick gypsum beds, which are nearly vertical and strike parallel with the axis of the syncline. In fact, it seems likely that subsidence is still active, for an area of an acre or more in the NE¼ sec. 22 and the NW¼ sec. 23, T. 5 S., R. 90 W., shows much evidence of slope creep and minor sinking which has affected the soil, sod, and brush.

PLEISTOCENE

GLACIAL DEPOSITS

Glacial deposits are widely distributed in the map area. The largest areas are on the plateau and in the wide valleys in the northern third of the area. The large irregular area of glacial deposits shown east of the basalt in the northeastern part of the map area, and those shown in T. 6 S., R. 87 W., may include deposits that are questionably glacial. Undoubted moraines are present for a considerable distance east of the rim of the plateau, but in the easternmost part of the area shown on the map as "glacial" a heterogeneous mixture of boulders, sand, and gravel is widespread. It may represent outwash deposits spread eastward by glacial streams.

Moraines extend down the northward-flowing streams on and beyond the north margin of the map area. Moraines extend to within a few miles of the Colorado River in the valleys of the southward-flowing main streams, and on Canyon Creek they extend nearly to the Colorado River. Many bare rock surfaces, especially those on granite and the Sawatch Quartzite, in the upper parts of the main stream valleys, including the South Fork of the White River, Kaiser, No Name, and Grizzly Creeks, contain striae, grooves, chatter marks, and other features formed by the valley glaciers as they moved downstream.

G. M. Richmond (written communication, 1951), who visited the area, reported that deposits of at least 1 pre-Wisconsin stage, 4 Wisconsin substages, and 2 post-Wisconsin advances are represented in the area. For example, a till having a pre-Wisconsin soil profile is exposed in sec. 15, T. 4 S., R. 88 W. (unsurveyed), in Crane Park. The soil profiles and other features of the two end moraines on Canyon Creek, according to Richmond, suggest that they represent deposits of the first and second substages of Wisconsin glaciation. General features of two prominent moraines on Sweetwater Creek below Sweetwater Lake suggest that they also probably represent the first two substages of Wisconsin glaciation. A moraine a short distance downstream from Trappers Lake near the north margin of the area and the large moraine at the lower end of the lake may represent the third and fourth substages, respectively, of Wisconsin glaciation, according to Richmond. Although he had examined the area only from aerial photographs, Richmond suggested that the small moraines enclosing lakes at altitudes of about 11,000 feet in the Flat Tops represent post-Wisconsin advances of Recent age.

PLEISTOCENE TO RECENT

RIVER TERRACES AND HIGH-LEVEL GRAVEL AND ALLUVIUM

Coarse cobbles and some boulders, evidently stream laid, were observed on a few high ridges at altitudes ranging from 1,500 to 2,700 feet above the Colorado River. Most cobbles and boulders are composed of tan quartzite and were probably derived from the Sawatch Quartzite.

Well-preserved gravel terraces are present 100 to 150 feet above the Colorado River on the north side of the river, from ¾ to 2 miles northwest of Glenwood Springs. Other terraces are present about 200 feet above the Colorado River on the north side of the river in the south part of T. 5 S., R. 90 W. In this category are alluvial fans including those southeast of the Colorado River in sec. 12, T. 5 S., R. 87 W., and the compound landslide and fan half a mile northwest of Glenwood Springs.

A relatively broad belt of alluvium is present at places 15 to 20 feet above stream bed along the Colorado, Roaring Fork, and Eagle Rivers. It is present in the eastern part of the map area, for several miles west and south of Glenwood Springs, and near the west margin of the map area. The Colorado, Eagle, and Roaring Fork Rivers are flowing on boulder-strewn beds.

TRAVERTINE

A deposit of travertine, which ranges in thickness from 3 to 40 feet, is present about 1 mile northwest of Glenwood Springs. It overlies a terrace of Colorado River cobbles where exposed along the road that passes northwestward in the NW¼ sec. 5, T. 6 S., R. 89 W.

SLIDE ROCK

A belt of slide rock, or talus, several hundred feet wide is present at many places at the foot of the cliffs of basalt at the edges of the Flat Tops and at the north and east edges of the large basalt-capped mesa in the southeastern part of the map area. The slide rock is composed of fresh, sharp-edged blocks of basalt that have undoubtedly broken from the high cliffs above. The two almost circular patches of basalt, each about one-fourth mile in diameter, shown on plate 1 in secs. 12 and 13, T. 6 S., R. 87 W., undoubtedly became detached from the main body of basalt and slid eastward and slightly downward, but remained as unbroken masses.

VOLCANIC ASH

A deposit of white volcanic ash, which consists of fine to very fine angular fragments of glass, is present one-fourth mile north of U.S. Highway 6 at the east end of Glenwood Canyon in the SE ¼ sec. 11, T. 5 S., R. 87 W. This deposit is present in only a few acres and attains a thickness of 25 feet or more. Concerning the fragments of glass, H. A. Powers reports (written communications, 1959):

> The glass shards of the ash are similar in shape to the shards of the Pearlette ash (Swineford and Frye, 1946), and the phenocryst assemblage is ferroaugite dominant over chevkinite, fayalite, brown hornblende and two kinds of zircon, similar to ash in Pine Valley, Utah, in the La Sal Mountains, Utah, near Golden, Colorado, and in the type areas of the Pearlette ash (Powers and others, 1958). Chemical composition of the glass shards has not been determined.

IGNEOUS ROCKS

MIOCENE(?) TO PLEISTOCENE

BASALT FLOWS

Basalt with a total thickness of about 1,000 feet and composed of 6 to 10 or more flows forms the rock of the Flat Tops in the northern one-third of the area. Beds of volcanic ash separate the flows in places, but commonly the flows appear to be in contact. Several local vents through which the lava issued include Sheep Mountain, Shingle and Trappers Peaks, Marvine Mountain, and two unnamed peaks 1¼ miles southeast of Marvine Lakes.

Basalt occupies a large area south of the Colorado River, in the southeastern part of the map area. The upper surface of the basalt here is at an altitude of about 9,000 feet, whereas in the Flat Tops it is about 11,500 feet. Several local vents, including Buck Point in sec. 21, T. 6 S., R. 87 W., and a small peak 3 miles west of it, apparently were sources of the flows present here. It is noteworthy that the lava beds slope at an appreciable angle southwestward and southward toward a topographic basin in secs. 14 and 15, T. 6 S., R. 87 W. The basin may have been formed in part prior to the outpouring of the lava as a result of collapse into a cavern in the underlying gypsum beds of the Paradox Formation. Other flows, lying at altitudes of about 7,000 to 7,500 feet, are present on the uplands on both sides of Roaring Fork River south of Glenwood Springs. The lava extends from a relatively flat plateau a considerable distance down the slope toward the Roaring Fork in the NW¼ sec. 17, 1½ miles south of west of Glenwood Springs, and in the S½ sec. 28, T. 6 S., R. 89 W., 2½ miles south of Glenwood Springs, which suggests that it was extruded following part of the cutting of Roaring Fork valley. It is probable that the flows lying west of Roaring Fork issued from Sunlight Peak, which is 8 miles southwest of Glenwood Springs and 5½ miles south of the map area.

Willow Peak, in sec. 32, T. 4 S., R. 87 W., is a cinder cone that is younger than the basalt flows just described, for the portion of the narrow tongue of lava extending eastward for 3½ miles from the cone fills the bottom of a narrow gulch in soft shale in secs. 35 and 36. Near the east end of the lava tongue the gulch has been cut to a level only about 25 feet lower than the base of the lava. An even younger small flow occupies a portion of the valley of Eagle River near Dotsero; it is reported to be younger than the Wisconsin stage of the Pleistocene (Landon, 1933). The flow apparently issued from a crevice in the northern part of the SE¼ sec. 33, T. 4 S., R. 86 W., about a mile north of Eagle River, and flowed southwestward down a small gulch to the Eagle River valley; lava still clings to the sides of the gulch. A large sinkhole immediately north of the dike that marks the old vent is erroneously referred to locally as the Dotsero Crater. A thick deposit of volcanic cinders covers an area of several acres near the lava vent.

Except for the flow at Dotsero, little is known about the age of the basalt flows in the map area. The age of the flows in the Flat Tops is generally designated as possibly Miocene, but evidence for their age is meager. It is not unlikely that the map area has experienced volcanism intermittently from Miocene(?) to late Pleistocene time.

BASALT SILLS

Two large sills of basalt are present in the Mancos Shale in T. 1 S., R. 86 W. The rock of Porphyry Mountain on Red Dirt Creek, and another mass 1 mile to the southwest, in T. 3 S., R. 86 W., is a light-gray igneous rock characterized by the presence of fragments of igneous and sedimentary rocks. It was tentatively identified in the field by J. D. Vine, who mapped the area, as quartz latite. Sugarloaf Mountain, a plug in sec. 5, T. 3 S., R. 86 W., and the broad dike extending southeastward from it, as well as the three masses of igneous rock one-half mile northwest and 2 miles west of it, seem to represent intrusions of quartz monzonite, according to Vine.

STRUCTURE

GENERAL STRUCTURAL FEATURES

The general attitude of the rocks in the map area is shown on the map (pl. 1) by structure contours drawn at intervals of 500 feet on the top of the Leadville Limestone. The map was prepared by superimposing the topographic map on the geologic map and estimating the altitude of the Leadville limestone at the several formation contacts. Inasmuch as the topographic map was made by plane-table sketching for publication at a scale of one-half inch to the mile and the geologic map was drawn from aerial photographs on a scale of 2 inches to the mile, the two maps do not coincide in detail.

Structurally the part of the White River uplift lying within the map area is a broad dome that is somewhat elongate northwestward. A large area in the western part of the uplift is as yet unmapped by the Geological Survey. Upon completion of the mapping there, the general shape of the uplift may be found to be different from that suggested by the present map. The rocks on parts of the southwest flank of the uplift in the map area dip steeply, at 50° to 90°, and locally are overturned. The rocks on the west flank, within the map area, and on the north, east, and southeast flanks dip moderately, at 8° to 10° for the most part. In an area about 15 miles wide by 18 miles long, just west of the center of the map area, the rocks have only gentle dips. This area constitutes the broad top of the dome. It is noteworthy that at least on the southwest flank of the uplift most of the structural relief is found in an area only a few miles wide.

Contour lines were not drawn in the vicinity of the Flat Tops in the northernmost part of the map area where lava flows obscure the sedimentary rocks. Moreover, the main orogeny that produced the White River uplift and the local faults and folds occurred prior to the outpouring of the lava that formed the Flat Tops.

JOINTS

All brittle beds in the map area are jointed. Commonly two sets of joints are present and their trends are at approximate right angles to each other. Joints are well developed and exposed on many parts of the broad plateau part of the area where broad areas of bare rock exposures are common.

FAULTS

The uplift, particularly in the western half of the map area, is characterized by many faults, both reverse and tension, many of which trend slightly north of west. The most pronounced of the faults, those that have the greatest structural effect, are a few thrust faults, including the Blair Mountain, Spring Creek, Dolan Gulch, Storm King, Grizzly Creek, and Red Table Mountain faults. The planes of the thrust faults in the south half of the area dip northward and in the northwestern part of the area dip westward. The few data available suggest that the planes of some of these faults dip at angles of less than 10°. The vertical displacement on a few of these faults ranges from 1,200 to 1,500 feet. The Red Table Mountain fault, however, near the southeast corner of the map area has a displacement of several thousand feet. The map area, especially the parts between the prominent thrust faults, contains many normal faults, most of which trend northwestward. The vertical displacement of the rocks along most of these faults ranges from less than 50 to about 400 feet, and the downdropped block is on the north. It seems reasonable to conclude that these normal faults formed at the time or shortly after the time of elevation of the blocks between the thrust faults, and may represent adjustments by gravity of segments of these elevated blocks.

The Red Table Mountain fault near the southeast corner of the map area produces the largest displacement of any of the faults. The rocks southeast of the fault dip southwestward at an angle of 50° or more; they form a portion of the southwest flank of an anticline which occupies a large area, known as Red Table Mountain, southeast of the map area. The trace of this fault is clearly visible on aerial photographs for more than 10 miles southwest of the map area, and was traced on the ground in reconnaissance northeastward, thence eastward for more than 12 miles beyond the map area.

Figure 1 is a sketch map of Colorado showing the principal known reverse faults in the State. According to Lovering and Goddard (1950), the fault planes of the major reverse faults in northern Colorado dip eastward or northeastward, and the planes of such faults in southern Colorado dip southwestward. The trend of the faults in the White River uplift, which are shown within the ellipse on figure 1, about at right angles to the trend of the major faults elsewhere in State and the dip of the fault planes is northward or northwestward. No definite explanation for the change in trend of the major reverse faults of the White River uplift from those elsewhere in Colorado is apparent. The region of the White River uplift, however, lies between the northern half of the State, where the relative movement of faults seems to have been westward, and the southern half of the State, where the relative movement seems to have been eastward. Its position, therefore, suggests that the area of the White River uplift may have been subjected to torsion and that the resultant stresses, which produced the reverse faults, probably acted in a southerly to southeasterly direction.

It follows logically that the White River uplift itself is largely the result of nearly horizontal compressive stress acting in a general easterly to southeasterly direction. Moreover, the White River uplift is but one small unit in the Rocky Mountain system, whose linear belts of folds and thrust faults surely indicate compressive forces acting from the west. On the other hand, the general structure of the uplift is that of a broad flat dome that shows considerable elevation as a unit. Some geologists (G. R. Downs, oral communication, 1953) contend that the uplift is the result of vertical stresses acting from great depth and that the faults and steep dips on the flanks of the uplift are merely minor features associated with the vertical mass movement.

MINERAL DEPOSITS

Several thick beds of coal in the Mesaverde Group crop out in the southeastern extension of the Grand Hogback, known also as Coal Ridge, in the southwestern part of the map area. The coal is of bituminous rank and is generally considered to be noncoking, although early reports refer to it as semicoking coal.

The New Castle truck mine, near the west boundary of the map area on the south side of the Colorado River, is operating in the Wheeler and Allen coal beds. Many years ago the Wheeler coal bed, 18 feet thick, the D coal bed, 5 feet thick, and beds in the Keystone coal group were mined in the valley of South Canyon Creek, which is about 5 miles west of Glenwood Springs.

Several formations, including the Chaffee, Leadville, Belden, Paradox, Maroon, Entrada, Morrison, Dakota, Mancos, Mesaverde, and Wasatch, contain beds that are prospectively valuable as reservoirs for oil and gas in areas adjacent to the map area. Beds of sandstone in the Mesaverde Group yield gas about 10 miles southwest of the map area. An interesting possible reservoir is a tongue of the Weber Sandstone, 100 to 200 feet thick, in the Maroon Formation. It crops out intermittently from the south boundary of the map area, 3 miles south of Glenwood Springs, northwestward across the southwestern part of the area, thence many miles farther northwest. The sandstone is impregnated with a black organic substance reported to be a petroleum residue. The Weber Sandstone is the principal oil reservoir in the Rangely oil field, 65 miles northwest of the map area, and in other oil fields in the region.

The presence of beds of fossiliferous marine limestone in the upper part of the Paradox Formation in the westernmost part of the map area and the absence of such limestone beds in the eastern part suggest the presence in the subsurface of porous bioherms or reefs a short distance west and northwest of the map area. The condition there may be comparable to that in southeastern Utah, where oil-bearing carbonates occur on the southwest flank of the Paradox salt basin.

Considerable prospecting for uranium has been carried on in recent years in the Morrison Formation in the southwestern part of the area.

Scores of shafts and prospect tunnels in the Leadville Limestone, particularly north of the Colorado River near the Denver and Rio Grande Western Railroad between Deep Creek and Grizzly Creek, bear evidence of a very active search for lead and zinc 70 or more years ago and an intermittent and small-scale search from that time until the present.

A rock quarry has been operated for many years in the Leadville Limestone on U.S. Highway 6, half a mile northwest of the residential part of Glenwood Springs. Part of the rock is shipped to sugar refineries, and part is processed by a plant at Glenwood Springs.

Many somewhat mineralized hot springs issue from river gravels at Glenwood Springs, where water from the largest spring is used to fill a large swimming pool. Caverns in the Leadville Limestone that are filled with natural steam are used for steam baths. A spring of relatively hot water issues from the Dakota Sandstone in South Canyon Creek valley, 5 miles west of Glenwood Springs. Although no specific data were collected concerning an explanation for the hot springs, the evidence of volcanism as recent as Pleistocene within the map area suggests the possibility that intrusive bodies may be present in the subsurface in the vicinity of the hot springs. It is noteworthy, also, that if the Storm King thrust fault were hypothetically projected southeastward beneath the alluvium of the Colorado River valley, it would pass beneath the hot springs at Glenwood Springs. A sample of water, whose temperature was 122° F. was collected June 19, 1955, by S. W. Lohman, from the Azure-Yampah spring at the Glenwood Springs swimming pool. Analysis by the Geological Survey laboratory at Denver, Colo., shows total dissolved solids of 19,100 parts per million and the principal chemical components as follows:

	ppm		ppm
Calcium	481	Bicarbonate	734
Magnesium	89	Sulfate	1, 130
Sodium	6, 690	Chloride	10, 100
Potassium	162		

The uncommonly large content of sodium chloride and magnesium sulfate suggests that the water passes through evaporite beds, salt and gypsum, of the Paradox Formation on its passage through the rocks.

Travertine suitable for building purposes is present in an area of 80 acres or more about 1 mile northwest of Glenwood Springs. The thickness of the travertine ranges from 3 to 40 feet.

Thick beds of gypsum in the Paradox Formation are present in many parts of the map area. Only on Blowout hill, where a stratigraphic section of the formation was measured, is the gypsum in place in beds; elsewhere it is contorted. The thickness of the thickest beds of gypsum on Blowout hill is, in ascending order, 160 feet, 75 feet, 150 feet, and 65 feet.

Volcanic cinders are being quarried from the cinder cone north of U.S. Highway 6 and the Eagle River, known locally as Dotsero Crater, in sec. 33, T. 4 S., R. 86 W. The cinders are trucked to a plant near Dotsero on the highway and are there manufactured into cinderblocks, used chiefly in the building industry. Another large deposit of volcanic cinders, composed generally of much coarser material than is present on Dotsero Crater, is in sec. 4, T. 7 S., R. 87 W., 2 miles south of the south boundary of the map area. Here a quarry is operated to obtain cinders for road surfacing in the general vicinity of the quarry.

The deposit of white volcanic ash, one-fourth mile north of U.S. Highway 6 at the east end of Glenwood Canyon and discussed previously, is present in only a few acres and attains a thickness of 25 feet or more. Likely its principal use would be as an ingredient for scouring powder.

LITERATURE CITED

Bass, N. W., 1956, Geology of the White River uplift in northwestern Colorado [summary]: Tulsa Geol. Soc. Digest, v. 24, p. 67–69.

———— 1958, Pennsylvanian and Permian rocks in the southern half of the White River uplift, Colorado, *in* Rocky Mountain Assoc. Geologists, Symposium on Pennsylvanian rocks of Colorado and adjacent areas, p. 91–94.

Bass, N. W., and Northrop, S. A., 1950, South Canyon Creek dolomite member, a unit of Phosphoria age in Maroon formation near Glenwood Springs, Colorado: Am. Assoc. Petroleum Geologists Bull., v. 34, p. 1540–1551.

———— 1953, Dotsero and Manitou formations, White River Plateau, Colorado, with special reference to Clinetop algal limestone member of Dotsero formation: Am. Assoc. Petroleum Geologists Bull., v. 37, p. 889–912.

———— 1955, Lower Paleozoic rocks of the White River uplift, Colorado, *in* Intermountain Assoc. Petroleum Geologists, Guidebook, 6th Ann. Field Conf. 1955, p. 3–9.

Bassett, C. F., 1938, Graptolites from Cambrian strata in Glenwood Canyon of the Colorado [abs.]: Geol. Soc. America Proc. 1937, p. 304–305.

———— 1939, Paleozoic section in the vicinity of Dotsero, Colorado: Geol. Soc. America Bull., v. 50, p. 1851–1865.

Bissell, H. J., and Childs, O. E., 1958, The Weber Formation of Utah and Colorado, *in* Rocky Mountain Assoc. Geologists, Symposium on Pennsylvanian rocks of Colorado and adjacent areas, pl. 1.

Bolyard, D. W., 1959, Pennsylvanian and Permian stratigraphy in Sangre de Cristo Mountains between La Veta Pass and Westcliffe, Colorado: Am. Assoc. Petroleum Geologists Bull., v. 43, p. 1896–1939.

Brill, K. G., Jr., 1942, Late Paleozoic stratigraphy of Gore area, Colorado: Am. Assoc. Petroleum Geologists Bull., v. 26, p. 1375–1397.

———— 1944, Late Paleozoic stratigraphy, west-central and northwestern Colorado: Geol. Soc. America Bull., v. 55, p. 621–655.

———— 1952, Stratigraphy in the Permo-Pennsylvanian zeugogeosyncline of Colorado and northern New Mexico: Geol. Soc. America Bull., v. 63, p. 809–880.

Bryant, W. L., and Johnson, J. H., 1936, Upper Devonian fish from Colorado: Jour. Paleontology, v. 10, no. 7, p. 656–659.

Cloud, P. E., Jr., and Barnes, V. E., 1948, Ellenburger group of central Texas: Texas Univ. Bur. Econ. Geology Pub. 4621, 473 p.

Cobban, W. A., and Reeside, J. B., Jr., 1951, Lower Cretaceous ammonites in Colorado, Wyoming, and Montana: Am. Assoc. Petroleum Geologists Bull., v. 35, p. 1892–1893.

Cooper, G. A., 1954, Unusual Devonian brachiopods: Jour. Paleontology, v. 28, p. 325–332.

Crickmay, C. H., 1952, Discrimination of late Upper Devonian: Jour. Paleontology, v. 26, p. 585–609.

Cross, C. W., Howe, E., and Ransome, F. L., 1905, Description of the Silverton quadrangle [Colo.]: U.S. Geol. Survey Geol. Atlas, Folio 120.

Denison, R. H., 1951, Late Devonian fresh-water fishes from the western United States: Fieldiana Geology, v. 11, p. 221–261.

Donnell, J. R., 1954, Tongue of Weber sandstone in Maroon formation near Carbondale and Redstone, northwestern Colorado: Am. Assoc. Petroleum Geologists Bull., v. 38, no. 8, p. 1817–1821.

Eastman, C. R., 1904, On Upper Devonian fish remains from Colorado: Am. Jour. Sci., 4th ser., v. 18, p. 253–260.

———— 1917, Fossil fishes in the collection of the United States National Museum: U.S. Natl. Mus. Proc., v. 52, p. 235–304.

Elias, M. K., 1957, Late Mississippian fauna from the Redoak Hollow formation of southern Oklahoma; pt. 2—Brachiopoda: Jour. Paleontology, v. 31, p. 487–527.

Fitzsimmons, J. P., Armstrong, A. K., and Gordon, Mackenzie, Jr., 1956, Arroyo Penasco formation, Mississippian, north-central New Mexico: Am. Assoc. Petroleum Geologists Bull., v. 40, p. 1935–1944.

Gehrig, J. L., 1958, Middle Pennsylvanian brachiopods from the Mud Springs Mountains and Derry Hills, New Mexico: New Mexico Bur. Mines and Mineral Res. Mem. 3, 24 p.

Girty, G. H., 1903, The Carboniferous formations and faunas of Colorado: U.S. Geol. Survey Prof. Paper 16, 546 p.

Hallgarth, W. E., 1959, Stratigraphy of Paleozoic rocks in northwestern Colorado: U.S. Geol. Survey Oil and Gas Inv. Chart OC–59.

Henbest, L. G., 1958, Significance of karst terrane and residuum in Upper Mississippian and Lower Pennsylvanian rocks, Rocky Mountain region, *in* Wyoming Geol. Assoc., Guidebook 13th Ann. Field Conf. 1958, p. 36–38.

Johnson, J. H., 1945, Calcareous algae of the upper Leadville limestone near Glenwood Springs, Colorado: Geol. Soc. America Bull., v. 56, p. 829–847.

Kindle, E. M., 1909, The Devonian fauna of the Ouray limestone: U.S. Geol. Survey Bull. 391, 60 p.

Landon, R. E., 1933, Date of recent volcanism in Colorado: Am. Jour. Sci., 5th ser., v. 25, p. 20–24.

Langenheim, R. L., Jr., 1952, Pennsylvanian and Permian stratigraphy in Crested Butte quadrangle, Gunnison County, Colorado: Am. Assoc. Petroleum Geologists Bull., v. 36, p. 543–574.

Laudon, L. R., and Bowsher, A. L., 1941, Mississippian formations of Sacramento Mountains, New Mexico: Am. Assoc. Petroleum Geologists Bull., v. 25, no. 12, p. 2107–2160.

Lovering, T. S., and Goddard, E. N., 1950, Geology and ore deposits of the Front Range, Colorado: U.S. Geol. Survey Prof. Paper 223, 319 p.

MacQuown, W. C., Jr., 1945, Structure of the White River Plateau near Glenwood Springs, Colorado: Geol. Soc. America Bull., v. 56, p. 877–892.

Maher, J. C., 1950, Pre-Pennsylvanian rocks along the Front Range of Colorado: U.S. Geol. Survey Oil and Gas Inv. Prelim. Chart 39.

Merrill, W. M., and Winar, R. M., 1958, Molas and associated formations in San Juan Basin—Needle Mountains area, southwestern Colorado: Am. Assoc. Petroleum Geologists Bull., v. 42, p. 2107–2132.

Moore, R. C., and others, 1944, Correlation of Pennsylvanian formations of North America: Geol. Soc. America Bull., v. 55, p. 657–706.

Muir-Wood, Helen, and Cooper, G. A., 1960, Morphology, classification, and life habits of the Productoidea (Brachiopoda): Geol. Soc. America Mem. 81, 448 p.

Peck, R. E., 1957, North American Mesozoic Charophyta: U.S. Geol. Survey Prof. Paper 294-A, p. 1–44.

Powers, H. A., Young, E. J., and Barnett, P. R., 1958, Possible extension into Idaho, Nevada, and Utah of the Pearlette ash of Meade County, Kansas [abs.]: Geol. Soc. America Bull., v. 69, p. 1631.

Schuchert, Charles, 1918, On the Carboniferous of the Grand Canyon of Arizona: Am. Jour. Sci., 4th ser., v. 45, p. 347–361.

Stainbrook, M. A., 1947, Brachiopoda of the Percha shale of New Mexico and Arizona: Jour. Paleontology, v. 21, p. 297–328.

Stensiö, E. A., 1948, On the Placodermi of the Upper Devonian of East Greenland, 2, Antiar chi, subfamily Bothriolepinae: Meddelelser om Grønland, Bind 139, 622 p.

Swineford, Ada, and Frye, J. C., 1946, Petrographic comparison of Pliocene and Pleistocene volcanic ash from western Kansas: Kansas State Geol. Survey Bull. 64, p. 1–32.

Thomas, C. R., McCann, F. T., and Raman, N. D., 1945, Mesozoic and Paleozoic stratigraphy in northwestern Colorado and northeastern Utah: U. S. Geol. Survey Oil and Gas Inv. Prelim. Chart 16.

Thompson, M. L., 1942, Pennsylvanian system in New Mexico: New Mexico Bur. Mines and Mineral Resources Bull. 17, 90 p.

———— 1945, Pennsylvanian rocks and fusulinids of east Utah and northwest Colorado correlated with Kansas section: Kansas State Geol. Survey Bull. 60, pt. 2, p. 17–84.

Tweto, O. L., 1949, Stratigraphy of the Pando area, Eagle County, Colorado: Colorado Sci. Soc. Proc., v. 15, p. 149–235.

Watts, L. F., and Dean, A. P., 1949, White River National Forest, Colorado: U.S. Forest Service Map.

Wengerd, S. A., and Strickland, J. W., 1954, Pennsylvanian stratigraphy of Paradox salt basin, Four Corners region, Colorado and Utah: Am. Assoc. Petroleum Geologists Bull., v. 38, p. 2157–2199.

Wood, G. H., Jr., and Northrop, S. A., 1946, Geology of Nacimiento Mountains, San Pedro Mountain, and adjacent plateaus in parts of Sandoval and Rio Arriba Counties, New Mexico: U.S. Geol. Survey Oil and Gas Inv. Prelim. Map 57.

INDEX

○

GOLD RUSH BOOKS

OREGON, USA

www.GoldMiningBooks.com

Books On Mining

Visit: www.goldminingbooks.com to order your copies or ask your favorite book seller to offer them.

Mining Books by Kerby Jackson

Gold Dust: Stories From Oregon's Mining Years - Oregon mining historian and prospector, Kerby Jackson, brings you a treasure trove of seventeen stories on Southern Oregon's rich history of gold prospecting, the prospectors and their discoveries, and the breathtaking areas they settled in and made homes. 5" X 8", 98 ppgs. Retail Price: $11.99

The Golden Trail: More Stories From Oregon's Mining Years - In his follow-up to "Gold Dust: Stories of Oregon's Mining Years", this time around, Jackson brings us twelve tales from Oregon's Gold Rush, including the story about the first gold strike on Canyon Creek in Grant County, about the old timers who found gold by the pail full at the Victor Mine near Galice, how Iradel Bray discovered a rich ledge of gold on the Coquille River during the height of the Rogue River War, a tale of two elderly miners on the hunt for a lost mine in the Cascade Mountains, details about the discovery of the famous Armstrong Nugget and others. 5" X 8", 70 ppgs. Retail Price: $10.99

Oregon Mining Books

Geology and Mineral Resources of Josephine County, Oregon - Unavailable since the 1970's, this important publication was originally compiled by the Oregon Department of Geology and Mineral Industries and includes important details on the economic geology and mineral resources of this important mining area in South Western Oregon. Included are notes on the history, geology and development of important mines, as well as insights into the mining of gold, copper, nickel, limestone, chromium and other minerals found in large quantities in Josephine County, Oregon. 8.5" X 11", 54 ppgs. Retail Price: $9.99

Mines and Prospects of the Mount Reuben Mining District - Unavailable since 1947, this important publication was originally compiled by geologist Elton Youngberg of the Oregon Department of Geology and Mineral Industries and includes detailed descriptions, histories and the geology of the Mount Reuben Mining District in Josephine County, Oregon. Included are notes on the history, geology, development and assay statistics, as well as underground maps of all the major mines and prospects in the vicinity of this much neglected mining district. 8.5" X 11", 48 ppgs. Retail Price: $9.99

The Granite Mining District - Notes on the history, geology and development of important mines in the well known Granite Mining District which is located in Grant County, Oregon. Some of the mines discussed include the Ajax, Blue Ribbon, Buffalo, Continental, Cougar-Independence, Magnolia, New York, Standard and the Tillicum. Also included are many rare maps pertaining to the mines in the area. 8.5" X 11", 48 ppgs. Retail Price: $9.99

Ore Deposits of the Takilma and Waldo Mining Districts of Josephine County, Oregon - The Waldo and Takilma mining districts are most notable for the fact that the earliest large scale mining of placer gold and copper in Oregon took place in these two areas. Included are details about some of the earliest large gold mines in the state such as the Llano de Oro, High Gravel, Cameron, Platerica, Deep Gravel and others, as well as copper mines such as the famous Queen of Bronze mine, the Waldo, Lily and Cowboy mines. This volume also includes six maps and 20 original illustrations. 8.5" X 11", 74 ppgs. Retail Price: $9.99

Metal Mines of Douglas, Coos and Curry Counties, Oregon - Oregon mining historian Kerby Jackson introduces us to a classic work on Oregon's mining history in this important re-issue of Bulletin 14C Volume 1, otherwise known as the Douglas, Coos & Curry Counties, Oregon Metal Mines Handbook. Unavailable since 1940, this important publication was originally compiled by the Oregon Department of Geology and Mineral Industries includes detailed descriptions, histories and the geology of over 250 metallic mineral mines and prospects in this rugged area of South West Oregon. 8.5" X 11", 158 ppgs. Retail Price: $19.99

Metal Mines of Jackson County, Oregon - Unavailable since 1943, this important publication was originally compiled by the Oregon Department of Geology and Mineral Industries includes detailed descriptions, histories and the geology of over 450 metallic mineral mines and prospects in Jackson County, Oregon. Included are such famous gold mining areas as Gold Hill, Jacksonville, Sterling and the Upper Applegate. **8.5" X 11", 220 ppgs. Retail Price: $24.99**

Metal Mines of Josephine County, Oregon - Oregon mining historian Kerby Jackson introduces us to a classic work on Oregon's mining history in this important re-issue of Bulletin 14C, otherwise known as the Josephine County, Oregon Metal Mines Handbook. Unavailable since 1952, this important publication was originally compiled by the Oregon Department of Geology and Mineral Industries includes detailed descriptions, histories and the geology of over 500 metallic mineral mines and prospects in Josephine County, Oregon. **8.5" X 11", 250 ppgs. Retail Price: $24.99**

Metal Mines of North East Oregon - Oregon mining historian Kerby Jackson introduces us to a classic work on Oregon's mining history in this important re-issue of Bulletin 14A and 14B, otherwise known as the North East Oregon Metal Mines Handbook. Unavailable since 1941, this important publication was originally compiled by the Oregon Department of Geology and Mineral Industries and includes detailed descriptions, histories and the geology of over 750 metallic mineral mines and prospects in North Eastern Oregon. **8.5" X 11", 310 ppgs. Retail Price: $29.99**

Metal Mines of North West Oregon - Oregon mining historian Kerby Jackson introduces us to a classic work on Oregon's mining history in this important re-issue of Bulletin 14D, otherwise known as the North West Oregon Metal Mines Handbook. Unavailable since 1951, this important publication was originally compiled by the Oregon Department of Geology and Mineral Industries and includes detailed descriptions, histories and the geology of over 250 metallic mineral mines and prospects in North Western Oregon. **8.5" X 11", 182 ppgs. Retail Price: $19.99**

Mines and Prospects of Oregon - Mining historian Kerby Jackson introduces us to a classic mining work by the Oregon Bureau of Mines in this important re-issue of The Handbook of Mines and Prospects of Oregon. Unavailable since 1916, this publication includes important insights into hundreds of gold, silver, copper, coal, limestone and other mines that operated in the State of Oregon around the turn of the 19th Century. Included are not only geological details on early mines throughout Oregon, but also insights into their history, production, locations and in some cases, also included are rare maps of their underground workings. **8.5" X 11", 314 ppgs. Retail Price: $24.99**

Lode Gold of the Klamath Mountains of Northern California and South West Oregon
(See California Mining Books)

Mineral Resources of South West Oregon - Unavailable since 1914, this publication includes important insights into dozens of mines that once operated in South West Oregon, including the famous gold fields of Josephine and Jackson Counties, as well as the Coal Mines of Coos County. Included are not only geological details on early mines throughout South West Oregon, but also insights into their history, production and locations. **8.5" X 11", 154 ppgs. Retail Price: $11.99**

Chromite Mining in The Klamath Mountains of California and Oregon
(See California Mining Books)

Southern Oregon Mineral Wealth - Unavailable since 1904, this rare publication provides a unique snapshot into the mines that were operating in the area at the time. Included are not only geological details on early mines throughout South West Oregon, but also insights into their history, production and locations. Some of the mining areas include Grave Creek, Greenback, Wolf Creek, Jump Off Joe Creek, Granite Hill, Galice, Mount Reuben, Gold Hill, Galls Creek, Kane Creek, Sardine Creek, Birdseye Creek, Evans Creek, Foots Creek, Jacksonville, Ashland, the Applegate River, Waldo, Kerby and the Illinois River, Althouse and Sucker Creek, as well as insights into local copper mining and other topics. **8.5" X 11", 64 ppgs. Retail Price: $8.99**

Geology and Ore Deposits of the Takilma and Waldo Mining Districts - Unavailable since the 1933, this publication was originally compiled by the United States Geological Survey and includes details on gold and copper mining in the Takilma and Waldo Districts of Josephine County, Oregon. The Waldo and Takilma mining districts are most notable for the fact that the earliest large scale mining of placer gold and copper in Oregon took place in these two areas. Included in this report are details about some of the earliest large gold mines in the state such as the Llano de Oro, High Gravel, Cameron, Platerica, Deep Gravel and others, as well as copper mines such as the famous Queen of Bronze mine, the Waldo, Lily and Cowboy mines. In addition to geological examinations, insights are also provided into the production, day to day operations and early histories of these mines, as well as calculations of known mineral reserves in the area. This volume also includes six maps and 20 original illustrations. **8.5" X 11", 74 ppgs. Retail Price: $9.99**

Gold Mines of Oregon - Oregon mining historian Kerby Jackson introduces us to a classic work on Oregon's mining history in this important re-issue of Bulletin 61, otherwise known as "Gold and Silver In Oregon". Unavailable since 1968, this important publication was originally compiled by geologists Howard C. Brooks and Len Ramp of the Oregon Department of Geology and Mineral Industries and includes detailed descriptions, histories and the geology of over 450 gold mines Oregon. Included are notes on the history, geology and gold production statistics of all the major mining areas in Oregon including the Klamath Mountains, the Blue Mountains and the North Cascades. While gold is where you find it, as every miner knows, the path to success is to prospect for gold where it was previously found. 8.5" X 11", 344 ppgs. Retail Price: $24.99

Mines and Mineral Resources of Curry County Oregon - Originally published in 1916, this important publication on Oregon Mining has not been available for nearly a century. Included are rare insights into the history, production and locations of dozens of gold mines in Curry County, Oregon, as well as detailed information on important Oregon mining districts in that area such as those at Agness, Bald Face Creek, Mule Creek, Boulder Creek, China Diggings, Collier Creek, Elk River, Gold Beach, Rock Creek, Sixes River and elsewhere. Particular attention is especially paid to the famous beach gold deposits of this portion of the Oregon Coast. 8.5" X 11", 140 ppgs. Retail Price: $11.99

Chromite Mining in South West Oregon - Originally published in 1961, this important publication on Oregon Mining has not been available for nearly a century. Included are rare insights into the history, production and locations of nearly 300 chromite mines in South Western Oregon. 8.5" X 11", 184 ppgs. Retail Price: $14.99

Mineral Resources of Douglas County Oregon - Originally published in 1972, this important publication on Oregon Mining has not been available for nearly forty years. Included are rare insights into the geology, history, production and locations of numerous gold mines and other mining properties in Douglas County, Oregon. 8.5" X 11", 124 ppgs. Retail Price: $11.99

Mineral Resources of Coos County Oregon - Originally published in 1972, this important publication on Oregon Mining has not been available for nearly forty years. Included are rare insights into the geology, history, production and locations of numerous gold mines and other mining properties in Coos County, Oregon. 8.5" X 11", 100 ppgs. Retail Price: $11.99

Mineral Resources of Lane County Oregon - Originally published in 1938, this important publication on Oregon Mining has not been available for nearly seventy five years. Included are extremely rare insights into the geology and mines of Lane County, Oregon, in particular in the Bohemia, Blue River, Oakridge, Black Butte and Winberry Mining Districts. 8.5" X 11", 82 ppgs. Retail Price: $9.99

Mineral Resources of the Upper Chetco River of Oregon: Including the Kalmiopsis Wilderness - Originally published in 1975, this important publication on Oregon Mining has not been available for nearly forty years. Withdrawn under the 1872 Mining Act since 1984, real insight into the minerals resources and mines of the Upper Chetco River has long been unavailable due to the remoteness of the area. Despite this, the decades of battle between property owners and environmental extremists over the last private mining inholding in the area has continued to pique the interest of those interested in mining and other forms of natural resource use. Gold mining began in the area in the 1850's and has a rich history in this geographic area, even if the facts surrounding it are little known. Included are twenty two rare photographs, as well as insights into the Becca and Morning Mine, the Emmly Mine (also known as Emily Camp), the Frazier Mine, the Golden Dream or Higgins Mine, Hustis Mine, Peck Mine and others. 8.5" X 11", 64 ppgs. Retail Price: $8.99

Gold Dredging in Oregon - Originally published in 1939, this important publication on Oregon Mining has not been available for nearly seventy five years. Included are extremely rare insights into the history and day to day operations of the dragline and bucketline gold dredges that once worked the placer gold fields of South West and North East Oregon in decades gone by. Also included are details into the areas that were worked by gold dredges in Josephine, Jackson, Baker and Grant counties, as well as the economic factors that impacted this mining method. This volume also offers a unique look into the values of river bottom land in relation to both farming and mining, in how farm lands were mined, re-soiled and reclamated after the dredges worked them. Featured are hard to find maps of the gold dredge fields, as well as rare photographs from a bygone era. 8.5" X 11", 86 ppgs. Retail Price: $8.99

Quick Silver Mining in Oregon - Originally published in 1963, this important publication on Oregon Mining has not been available for over fifty years. This publication includes details into the history and production of Elemental Mercury or Quicksilver in the State of Oregon. 8.5" X 11", 238 ppgs. Retail Price: $15.99

Mines of the Greenhorn Mining District of Grant County Oregon - Originally published in 1948, this important publication on Oregon Mining has not been available for over sixty five years. In this publication are rare insights into the mines of the famous Greenhorn Mining District of Grant County, Oregon, especially the famous Morning Mine. Also included are details on the Tempest, Tiger, Bi-Metallic, Windsor, Psyche, Big Johnny, Snow Creek, Banzette and Paramount Mines, as well as prospects in the vicinities in the famous mining areas of Mormon Basin, Vinegar Basin and Desolation Creek. Included are hard to find mine maps and dozens of rare photographs from the bygone era of Grant County's rich mining history. 8.5" X 11", 72 ppgs. Retail Price: $9.99

Geology of the Wallowa Mountains of Oregon: Part I (Volume 1) - Originally published in 1938, this important publication on Oregon Mining has not been available for nearly seventy five years. Included are details on the geology of this unique portion of North Eastern Oregon. This is the first part of a two book series on the area. Accompanying the text are rare photographs and historic maps.**8.5" X 11", 92 ppgs. Retail Price: $9.99**

Geology of the Wallowa Mountains of Oregon: Part II (Volume 2) - Originally published in 1938, this important publication on Oregon Mining has not been available for nearly seventy five years. Included are details on the geology of this unique portion of North Eastern Oregon. This is the first part of a two book series on the area. Accompanying the text are rare photographs and historic maps.**8.5" X 11", 94 ppgs. Retail Price: $9.99**

Field Identification of Minerals For Oregon Prospectors - Originally published in 1940, this important publication on Oregon Mining has not been available for nearly seventy five years. Included in this volume is an easy system for testing and identifying a wide range of minerals that might be found by prospectors, geologists and rockhounds in the State of Oregon, as well as in other locales. Topics include how to put together your own field testing kit and how to conduct rudimentary tests in the field. This volume is written in a clear and concise way to make it useful even for beginners. **8.5" X 11", 158 ppgs. Retail Price: $14.99**

The Bohemia Mining District of Oregon - Originally published in 1900, this important publication on Oregon Mining has not been available for over a century. Included in this volume are important insights into the famous Bohemia Mining District of Oregon, including the histories and locations of important gold mines in the area such as the Ophir Mine, Clarence, Acturas, Peek-a-boo, White Swan, Combination Mine, the Musick Mine, The California, White Ghost, The Mystery, Wall Street, Vesuvius, Story, Lizzie Bullock, Delta, Elsie Dora, Golden Slipper, Broadway, Champion Mine, Knott, Noonday, Helena, White Wings, Riverside and others. Also included are notes on the nearby Blue River Mining District. **8.5" X 11", 58 ppgs. Retail Price: $9.99**

The Gold Fields of Eastern Oregon - Unavailable since 1900, this publication was originally compiled by the Baker City Chamber of Commerce Offering important insights into the gold mining history of Eastern Oregon, "The Gold Fields of Eastern Oregon" sheds a rare light on many of the gold mines that were operating at the turn of the 19th Century in Baker County and Grant County in North Eastern Oregon. Some of the areas featured include the Cable Cove District, Baisely-Elhorn, Granite, Red Boy, Bonanza, Susanville, Sparta, Virtue, Vaughn, Sumpter, Burnt River, Rye Valley and other mining districts. Included is basic information on not only many gold mines that are well known to those interested in Eastern Oregon mining history, but also many mines and prospects which have been mostly lost to the passage of time. Accompanying are numerous rare photos **8.5" X 11", 78 ppgs. Retail Price: $10.99**

Gold Mining in Eastern Oregon - Originally published in 1938, this important publication on Oregon Mining has not been available for over a century. Included in this volume are important insights into the famous mining districts of Eastern Oregon during the late 1930's. Particular attention is given to those gold mines with milling and concentrating facilities in the Greenhorn, Red Boy, Alamo, Bonanza, Granite, Cable Cove, Cracker Creek, Virtue, Keating, Medical Springs, Sanger, Sparta, Chicken Creek, Mormon Basin, Connor Creek, Cornucopia and the Bull Run Mining Districts. Some of the mines featured include the Ben Harrison, North Pole-Columbia, Highland Maxwell, Baisley-Elkhorn, White Swan, Balm Creek, Twin Baby, Gem of Sparta, New Deal, Gleason, Gifford-Johnson, Cornucopia, Record, Bull Run, Orion and others. Of particular interest are the mill flow sheets and descriptions of milling operations of these mines. **8.5" X 11", 68 ppgs. Retail Price: $8.99**

The Gold Belt of the Blue Mountains of Oregon - Originally published in 1901, this important publication on Oregon Mining has not been available for over a century. Included in this volume are rare insights into the gold deposits of the Blue Mountains of North East Oregon, including the history of their early discovery and early production. Extensive details are offered on this important mining area's mineralogy and economic geology, as well as insights into nearby gold placers, silver deposits and copper deposits. Featured are the Elkhorn and Rock Creek mining districts, the Pocahontas district, Auburn and Minersville districts, Sumpter and Cracker Creek, Cable Cove, the Camp Carson district, Granite, Alamo, Greenhorn, Robinsonville, the Upper Burnt River Valley and Bonanza districts, Susanville, Quartzburg, Canyon Creek, Virtue, the Copper Butte district, the North Powder River, Sparta, Eagle Creek, Cornucopia, Pine Creek, Lower Powder River, the Upper Snake River Canyon, Rye Valley, Lower Burnt River Valley, Mormon Basin, the Malheur and Clarks Creek districts, Sutton Creek and others. Of particular interest are important details on numerous gold mines and prospects in these mining districts, including their locations, histories, geology and other important information, as well as information on silver, copper and fire opal deposits. **8.5" X 11", 250 ppgs. Retail Price: $24.99**

Mining in the Cascades Range of Oregon - Originally published in 1938, this important publication on Oregon Mining has not been available for over seventy five years. Included in this volume are rare insights into the gold mines and other types of metal mines in the Cascades Mountain Range of Oregon. Some of the important mining areas covered include the famous Bohemia Mining District, the North Santiam Mining District, Quartzville Mining District, Blue River Mining District, Fall Creek Mining District, Oakridge District, Zinc District, Buzzard-Al Sarena District, Grand Cove, Climax District and Barron Mining District. Of particular interest are important details on over 100 mines and prospects in these mining districts, including their locations, histories, geology and other important information. **8.5" X 11", 170 ppgs. Retail Price: $14.99**

Beach Gold Placers of the Oregon Coast - Originally published in 1934, this important publication on Oregon Mining has not been available for over 80 years. Included in this volume are rare insights into the beach gold deposits of the State of Oregon, including their locations, occurance, composition and geology. Of particular interest is information on placer platinum in Oregon's rich beach deposits. Also included are the locations and other information on some famous Oregon beach mines, including the Pioneer, Eagle, Chickamin, Iowa and beach placer mines north of the mouth of the Rogue River. **8.5" X 11", 60 ppgs. Retail Price: $8.99**

Mineralogical Composition of the Sands of the Oregon Coast: From Coos Bay to the Columbia - Published in 1945, he text features hard to find information on the composition of the gold bearing black sands of the South West Oregon Coast, offering a unique insight to prospectors in search of Oregon's legendary beach gold. 104 ppgs, $9.99

Manganese Mining in Oregon - First released in 1942 and now out of print, this special reprint edition of "Manganese in Oregon" was originally published by the Oregon Department of Geology and Mineral Industries. The text features hard to find information on the mining of Manganese in Oregon, including details and maps of Oregon manganese mines and prospects. 108 ppgs, 9.99

Medford Oregon As A Mining Center - Written in 1912, this hard to find publication includes valuable insights into the mining history of South West Oregon. This small book contains interesting information on the gold, copper and mining industry in Southern Oregon as it existed just prior to World War One, shedding light on some of the important mines in the area. Included are rare photographs and vintage advertising of the day. 80 ppgs, 9.99

Mineral Resources of Curry County Oregon - First released in 1977 and now out of print, this special reprint edition of "Geology, Mineral Resources and Rock Materials of Curry County, Oregon" was originally published in cooperation of Curry County, Oregon and the Oregon Department of Geology and Mineral Industries. The text features hard to find information on not only the mining of gold and other metals in Curry County, but also aggregate mining in the area. 102 ppgs, 11.99

Origin of the Gold Bearing Black Sands of the Coast of South West Oregon - First released in 1943 and now out of print, this special reprint edition of "The Origin of the Black Sands of the South West Oregon Coast" was originally published by the Oregon Department of Geology and Mineral Industries. The text features hard to find information on the origin of the gold bearing black sands of the South West Oregon Coast, offering a unique insight to prospectors in search of Oregon's legendary beach gold. 52 ppgs, 8.99

South West Oregon Mining - Leading mining historian Kerby Jackson introduces us to six classic small mining publications on the Gold Mining Industry in Southern Oregon. This small book consists of a compilation of USGS J.S. Diller's "Mines of the Riddles Quadrangle", "The Rogue River Valley Coal Fields" and "Mineral Resources of the Grants Pass Quadrangle", the Grants Pass Commercial Club's rare publication "Mining in Josephine County, Oregon" and the USGS publication "The Distribution of Placer Gold in the Sixes River, South West Oregon". Also included is F.W. Libbey's legendary article on the Southern Oregon Mining Industry, "Lest We Forget", which appeared in the publication of the Oregon State Department of Geology and Mineral Industries in the early 1960's. This compilation offers a unique perspective on mining in South West Oregon and includes considerable information on mines in Josephine, Jackson and Coos Counties. 142 ppgs, 14.99

Geology and Mineral Resources of the Gasquet Quadrangle of California-Oregon - First published in 1953, it has been unavailable for over a century and sheds important light on the geological features and mineral resources of this portion of Northern California and Southern Oregon. 80 ppgs, 9.99

Idaho Mining Books

Gold in Idaho - Unavailable since the 1940's, this publication was originally compiled by the Idaho Bureau of Mines and includes details on gold mining in Idaho. Included is not only raw data on gold production in Idaho, but also valuable insight into where gold may be found in Idaho, as well as practical information on the gold bearing rocks and other geological features that will assist those looking for placer and lode gold in the State of Idaho. This volume also includes thirteen gold maps that greatly enhance the practical usability of the information contained in this small book detailing where to find gold in Idaho. **8.5" X 11", 72 ppgs. Retail Price: $9.99**

Geology of the Couer D'Alene Mining District of Idaho - Unavailable since 1961, this publication was originally compiled by the Idaho Bureau of Mines and Geology and includes details on the mining of gold, silver and other minerals in the famous Coeur D'Alene Mining District in Northern Idaho. Included are details on the early history of the Coeur D'Alene Mining District, local tectonic settings, ore deposit features, information on the mineral belts of the Osburn Fault, as well as detailed information on the famous Bunker Hill Mine, the Dayrock Mine, Galena Mine, Lucky Friday Mine and the infamous Sunshine Mine. This volume also includes sixteen hard to find maps. **8.5" X 11", 70 ppgs. Retail Price: $9.99**

The Gold Camps and Silver Cities of Idaho - Originally published in 1963, this important publication on Idaho Mining has not been available for nearly fifty years. Included are rare insights into the history of Idaho's Gold Rush, as well as the mad craze for silver in the Idaho Panhandle. Documented in fine detail are the early mining excitements at Boise Basin, at South Boise, in the Owyhees, at Deadwood, Long Valley, Stanley Basin and Robinson Bar, at Atlanta, on the famous Boise River, Volcano, Little Smokey, Banner, Boise Ridge, Hailey, Leesburg, Lemhi, Pearl, at South Mountain, Shoup and Ulysses, Yellow Jacket and Loon Creek. The story follows with the appearance of Chinese miners at the new mining camps on the Snake River, Black Pine, Yankee Fork, Bay Horse, Clayton, Heath, Seven Devils, Gibbonsville, Vienna and Sawtooth City. Also included are special sections on the Idaho Lead and Silver mines of the late 1800's, as well as the mining discoveries of the early 1900's that paved the way for Idaho's modern mining and mineral industry. Lavishly illustrated with rare historic photos, this volume provides a one of a kind documentary into Idaho's mining history that is sure to be enjoyed by not only modern miners and prospectors who still scour the hills in search of nature's treasures, but also those enjoy history and tromping through overgrown ghost towns and long abandoned mining camps. **8.5" X 11", 186 ppgs. Retail Price: $14.99**

Ore Deposits and Mining in North Western Custer County Idaho - Unavailable since 1913, this important publication was originally published by the Us Department of the Interior and has been unavailable for a century. Included are fine details on the geology, geography, gold placers and gold and silver bearing quartz veins of the mining region of North West Custer County, Idaho. Of particular interest is a rare look at the mines and prospects of the region, including those such as the Ramshorn Mine, SkyLark, Riverview, Excelsior, Beardsley, Pacific, Hoosier, Silver Brick, Forest Rose and dozens of others in the Bay Horse Mining District. Also covered are the mines of the Yankee Fork District such as the Lucky Boy, Badger, Black, Enterprise, Charles Dickens, Morrison, Golden Sunbeam, Montana, Golden Gate and others, as well as those in the Loon Mining District. **8.5" X 11", 126 ppgs. Retail Price: $12.99**

Gold Rush To Idaho - Unavailable since 1963, this important publication was originally published by the Idaho Bureau of Mines and has been unavailable for 50 years. "Gold Rush To Idaho" revisits the earliest years of the discovery of gold in Idaho Territory and introduces us to the conditions that the pioneer gold seekers met when they blazed a trail through the wilderness of Idaho's mountains and discovered the precious yellow metal at Oro Fino and Pierce. Subsequent rushes followed at places like Elk City, Newsome, Clearwater Station, Florence, Warrens and elsewhere. Of particular interest is a rare look at the hardships that the first miners in Idaho met with during their day to day existences and their attempts to bring law and order to their mining camps. **8.5" X 11", 88 ppgs. Retail Price: $9.99**

The Geology and Mines of Northern Idaho and North Western Montana - Unavailable since 1909, this important publication was originally published by the Us Department of the Interior and has been unavailable for a century. Included are fine details on the geology and geography of the mining regions of Northern Idaho and North Western Montana. Of particular interest is a rare look at the mines and prospects of the region, including those in the Pine Creek Mining District, Lake Pend Oreille district, Troy Mining District, Sylvanite District, Cabinet Mining District, Prospect Mining District and the Missoula Valley. Some of the mines featured include the Iron Mountain, Silver Butte, Snowshoe, Grouse Mountain Mine and others. **8.5" X 11", 142 ppgs. Retail Price: $12.99**

Mining in the Alturas Quadrangle of Blaine County Idaho - Unavailable since 1922, this important publication was originally published by the Idaho Bureau of Mines and has been unavailable for ninety years. Topics include the geology, rock formations and the formation of ore deposits in this important mining area of Idaho. Of particular focus is information on the local geology, quartz veins and ore deposits of this portion of Idaho. Included are hard to find details, including the descriptions and locations of numerous gold and silver mines in the area including the Silver King, Pilgrim, Columbia, Lone Jack, Sunbeam, Pride of the West, Lucky Boy, Scotia, Atlanta, Beaver-Bidwell and others mines and prospects. **8.5" X 11", 56 ppgs. Retail Price: $8.99**

Mining in Lemhi County Idaho - Originally published in 1913, this important book on Idaho Mining has not been available to miners for over a century. Included are rare insights into hundreds of gold, silver, copper and other mines in this famous Idaho mining area. Details include the locations, geology, history, production and other facts of the mines of this region, not only gold and silver hardrock mines, but also gold placer mines, lead-silver deposits, copper mines, cobalt-nickel deposits, tungsten and tin mines . It is lavishly illustrated with hard to find photos of the period and rare mining maps. Some of the vicinities featured include the Nicholia Mining District, Spring Mountain District, Texas District, Blue Wing District, Junction District, McDevitt District, Pratt Creek, Eldorado District, Kirtley Creek, Carmen Creek, Gibbonsville, Indian Creek, Mineral Hill District, Mackinaw, Eureka District, Blackbird District, YellowJacket District, Gravel Range District, Junction District, Parker Mountain and other mining districts. 8.5" X 11", 226 ppgs. Retail Price: $19.99

Mining in Shoshone County Idaho - First published in 1923, it has been unavailable for over a century and sheds important light on the mining history of Shoshone County, Idaho. Some of the topics include the history of mining in Shoshone County, a look at the local geology and ore characteristics of lead-silver deposits, zinc deposits, copper, antimony, gold and other minerals. Also included are insights into the history, production, characteristics and locations of numerous mines in the area. 198 ppgs, 15.99

Utah Mining Books

Fluorite in Utah - Unavailable since 1954, this publication was originally compiled by the USGS, State of Utah and U.S. Atomic Energy Commission and details the mining of fluorspar, also known as fluorite in the State of Utah. Included are details on the geology and history of fluorspar (fluorite) mining in Utah, including details on where this unique gem mineral may be found in the State of Utah. 8.5" X 11", 60 ppgs. Retail Price: $8.99

The Gold Hill Mining District of Utah - First published in 1935, it has been unavailable since those days and sheds important light on the mines, history and geology of Utah's Gold Hill Mining District. Included are rare insights into this important mining area, including the locations, histories and details of numerous mines. This volume is well illustrated with geological diagrams, as well as hard to find maps of some of the most important mines in this district. 202 ppgs., 19.99

The Mines, Miners and Minerals of Utah - First published in 1896, it has been unavailable since those days and sheds important light on the early mines and miners of Pioneer Utah, as well as the minerals which they won from the earth by laborious hard physical labor and sheer determination. Included are rare insights into the early mining history of Utah, as well details on hundreds of gold, silver and copper mines. 376 ppgs., 24.99

California Mining Books

The Tertiary Gravels of the Sierra Nevada of California - Mining historian Kerby Jackson introduces us to a classic mining work by Waldemar Lindgren in this important re-issue of The Tertiary Gravels of the Sierra Nevada of California. Unavailable since 1911, this publication includes details on the gold bearing ancient river channels of the famous Sierra Nevada region of California. 8.5" X 11", 282 ppgs. Retail Price: $19.99

The Mother Lode Mining Region of California - Unavailable since 1900, this publication includes details on the gold mines of California's famous Mother Lode gold mining area. Included are details on the geology, history and important gold mines of the region, as well as insights into historic mining methods, mine timbering, mining machinery, mining bell signals and other details on how these mines operated. Also included are insights into the gold mines of the California Mother Lode that were in operation during the first sixty years of California's mining history. 8.5" X 11", 176 ppgs. Retail Price: $14.99

Lode Gold of the Klamath Mountains of Northern California and South West Oregon - Unavailable since 1971, this publication was originally compiled by Preston E. Hotz and includes details on the lode mining districts of Oregon and California's Klamath Mountains. Included are details on the geology, history and important lode mines of the French Gulch, Deadwood, Whiskeytown, Shasta, Redding, Muletown, South Fork, Old Diggings, Dog Creek (Delta), Bully Choop (Indian Creek), Harrison Gulch, Hayfork, Minersville, Trinity Center, Canyon Creek, East Fork, New River, Denny, Liberty (Black Bear), Cecilville, Callahan, Yreka, Fort Jones and Happy Camp mining districts in California, as well as the Ashland, Rogue River, Applegate, Illinois River, Takilma, Greenback, Galice, Silver Peak, Myrtle Creek and Mule Creek districts of South Western Oregon. Also included are insights into the mineralization and other characteristics of this important.mining region. 8.5" X 11", 100 ppgs. Retail Price: $10.99

Mines and Mineral Resources of Shasta County, Siskiyou County, Trinity County: California - Unavailable since 1915, this publication was originally compiled by the California State Mining Bureau and includes details on the gold mines of this area of Northern California. Also included are insights into the mineralization and other characteristics of this important mining region, as well as the location of historic gold mines. 8.5" X 11", 204 ppgs. Retail Price: $19.99

Geology of the Yreka Quadrangle, Siskiyou County, California - Unavailable since 1977, this publication was originally compiled by Preston E. Hotz and includes details on the geology of the Yreka Quadrangle of Siskiyou County, California. Also included are insights into the mineralization and other characteristics of this important mining region. **8.5" X 11", 78 ppgs. Retail Price: $7.99**

Mines of San Diego and Imperial Counties, California - Originally published in 1914, this important publication on California Mining has not been available for a century. This publication includes important information on the early gold mines of San Diego and Imperial County, which were some of the first gold fields mined in California by early Spanish and Mexican miners before the 49ers came on the scene. Included are not only details on early mining methods in the area, production statistics and geological information, but also the location of the early gold mines that helped make California "The Golden State". Also included are details on the mining of other minerals such as silver, lead, zinc, manganese, tungsten, vanadium, asbestos, barite, borax, cement, clay, dolomite, fluospar, gem stones, graphite, marble, salines, petroleum, stronium, talc and others. **8.5" X 11", 116 ppgs. Retail Price: $12.99**

Mines of Sierra County, California - Unavailable since 1920, this publication was originally compiled by the California State Mining Bureau and includes details on the gold mines of Sierra County, California. Also included are insights into the mineralization and other characteristics of this important mining region, as well as the location of historic gold mines. **8.5" X 11", 156 ppgs. Retail Price: $19.99**

Mines of Plumas County, California - Unavailable since 1918, this publication was originally compiled by the California State Mining Bureau and includes details on the gold mines of Plumas County, California. Also included are insights into the mineralization and other characteristics of this important mining region, as well as the location of historic gold mines. **8.5" X 11", 200 ppgs. Retail Price: $19.99**

Mines of El Dorado, Placer, Sacramento and Yuba Counties, California - Originally published in 1917, this important publication on California Mining has not been available for nearly a century. This publication includes important information on the early gold mines of El Dorado County, Placer County, Sacramento County and Yuba County, which were some of the first gold fields mined by the Forty-Niners during the California Gold Rush. Included are not only details on early mining methods in the area, production statistics and geological information, but also the location of the early gold mines that helped make California "The Golden State". Also included are insights into the early mining of chrome, copper and other minerals in this important mining area. **8.5" X 11", 204 ppgs. Retail Price: $19.99**

Mines of Los Angeles, Orange and Riverside Counties, California - Originally published in 1917, this important publication on California Mining has not been available for a century. This publication includes important information on the early gold mines of Los Angeles County, Orange County and Riverside County, which were some of the first gold fields mined in California by early Spanish and Mexican miners before the 49ers came on the scene. Included are not only details on early mining methods in the area, production statistics and geological information, but also the location of the early gold mines that helped make California "The Golden State". **8.5" X 11", 146 ppgs. Retail Price: $12.99**

Mines of San Bernadino and Tulare Counties, California - Originally published in 1917, this important publication on California Mining has not been available for nearly a century. This publication includes important information on the early gold mines of San Bernadino and Tulare County, which were some of the first gold fields mined in California by early Spanish and Mexican miners before the 49ers came on the scene. Included are not only details on early mining methods in the area, production statistics and geological information, but also the location of the early gold mines that helped make California "The Golden State". Also included are details on the mining of other minerals such as copper, iron, lead, zinc, manganese, tungsten, vanadium, asbestos, barite, borax, cement, clay, dolomite, fluospar, gem stones, graphite, marble, salines, petroleum, stronium, talc and others. **8.5" X 11", 200 ppgs. Retail Price: $19.99**

Chromite Mining in The Klamath Mountains of California and Oregon - Unavailable since 1919, this publication was originally compiled by J.S. Diller of the United States Department of Geological Survey and includes details on the chromite mines of this area of Northern California and Southern Oregon. Also included are insights into the mineralization and other characteristics of this important mining region, as well as the location of historic mines. Also included are insights into chromite mining in Eastern Oregon and Montana. **8.5" X 11", 98 ppgs. Retail Price: $9.99**

Mines and Mining in Amador, Calaveras and Tuolumne Counties, California - Unavailable since 1915, this publication was originally compiled by William Tucker and includes details on the mines and mineral resources of this important California mining area. Included are details on the geology, history and important gold mines of the region, as well as insights into other local mineral resources such as asbestos, clay, copper, talc, limestone and others. Also included are insights into the mineralization and other characteristics of this important portion of California's Mother Lode mining region. **8.5" X 11", 198 ppgs. Retail Price: $14.99**

The Cerro Gordo Mining District of Inyo County California - Unavailable since 1963, this publication was originally compiled by the United States Department of Interior. Included are insights into the mineralization and other characteristics of this important mining region of Southern California. Topics include the mining of gold and silver in this important mining district in Inyo County, California, including details on the history, production and locations of the Cerro Gordo Mine, the Morning Star Mine, Estelle Tunnel, Charles Lease Tunnel, Ignacio, Hart, Crosscut Tunnel, Sunset, Upper Newtown, Newtown, Ella, Perseverance, Newsboy, Belmont and other silver and gold mines in the Cerro Gordo Mining District. This volume also includes important insights into the fossil record, geologic formations, faults and other aspects of economic geology in this California mining district. **8.5" X 11", 104 ppgs. Retail Price: $10.99**

Mining in Butte, Lassen, Modoc, Sutter and Tehama Counties of California - Unavailable since 1917, this publication was originally compiled by the United States Department of Interior. Included are insights into the mineralization and other characteristics of this important mining region of California. Topics include the mining of asbestos, chromite, gold, diamonds and manganese in Butte County, the mining of gold and copper in the Hayden Hill and Diamond Mountain mining districts of Lassen County, the mining of coal, salt, copper and gold in the High Grade and Winters mining districts of Modoc County, gold mining in Sutter County and the mining of gold, chromite, manganese and copper in Tehama County. This volume also includes the production records and locations of numerous mines in this important mining region. **8.5" X 11", 114 ppgs. Retail Price: $11.99**

Mines of Trinity County California - Originally published in 1965, this important publication on California Mining has not been available for nearly fifty years. This publication includes important information on mines and mining in Trinity County, California, as well insights into the mineralization and geology of this important mining area in Northern California. Included are extensive details on hardrock and placer gold mines and prospects, including charts showing the locations of these historic mines.. **8.5" X 11", 144 ppgs. Retail Price: $12.99**

Mines of Kern County California - Originally published in 1962, this important publication on California Mining has not been available for nearly fifty years. This publication includes important information on mines and mining in Kern County, California, as well insights into the mineralization and geology of this important mining area in California. Included are extensive details on hardrock and placer gold mines and prospects, including charts showing the locations of these historic mines. **8.5" X 11", 398 ppgs. Retail Price: $24.99**

Mines of Calaveras County California - Originally published in 1962, this important publication on California Mining has not been available for nearly fifty years. This publication includes important information on mines and mining in Calaveras County, California, as well insights into the mineralization and geology of this important mining area in Northern California. Included are extensive details on hardrock and placer gold mines and prospects, including charts showing the locations of these historic mines. **8.5" X 11", 236 ppgs. Retail Price: $19.99**

Lode Gold Mining in Grass Valley California - Unavailable since 1940, this publication was originally compiled by the United States Department of Interior. Included are insights into the gold mineralization and other characteristics of this important mining region of Nevada County, California. This volume also includes important insights into the geologic formations, faults and other aspects of economic geology in this California mining district. Of particular interest are the fine details on many hardrock gold mines in the area, including their locations, histories, development and mineralization. Some of the mines featured include the Gold Hill Mine, Massachusetts Hill, Boundary, Peabody, Golden Center, North Star, Omaha, Lone Jack, Homeward Bound, Hartery, Wisconsin, Allison Ranch, Phoenix, Kate Hayes, W.Y.O.D., Empire, Rich Hill, Daisy Hill, Orleans, Sultana, Centennial, Conlin, Ben Franklin, Crown Point and many others. **8.5" X 11", 148 ppgs. Retail Price: $12.99**

Lode Mining in the Alleghany District of Sierra County California - Unavailable since 1913, this publication was originally compiled by the United States Department of Interior. Included are insights into the mineralization and other characteristics of this important mining region of Sierra County. Included are details on the history, production and locations of numerous hardrock gold mines in this famous California area, including the Tightner Mine, Minnie D., Osceola, Eldorado, Twenty One, Sherman, Kenton, Oriental, Rainbow, Plumbago, Irelan, Gold Canyon, North Fork, Federal, Kate Hardy and others. This volume also includes important insights into the fossil record, geologic formations, faults and other aspects of economic geology in this California mining district. **8.5" X 11", 48 ppgs. Retail Price: $7.99**

Six Months In The Gold Mines During The California Gold Rush - Unavailable since 1850, this important work is a first hand account of one "49'ers" personal experience during the great California Gold Rush, shedding important light on one of the most exciting periods in the history of not only California, but also the world. Compiled from journals written between 1847 and 1849 by E. Gould Buffum, a native of New York, "Six Months In The Gold Mines During The California Gold Rush" offers a rare look into the day to day lives of the people who came to California to work in her gold mines when the state was still a great frontier. **8.5" X 11", 290 ppgs. Retail Price: $19.99**

Quartz Mines of the Grass Valley Mining District of California - Unavailable since 1867, this important publication has not been available since those days. This rare publication offers a short dissertation on the early hardrock mines in this important mining district in the California Mother Lode region between the 1850's and 1860's. Also included are hard to find details on the mineralization and locations of these mines, as well as how they were operated in those day. **8.5" X 11", 44 ppgs. Retail Price: $8.99**

Gold Rush on the Feather River - First published in 1924, this short publication by G.C. Mansfield sheds important light on the early history of gold mining on the Feather River. Included are rare insights into the first decade of gold mining and the early mining camps of the Feather River during the 1850's. 64 ppgs., 9.99

The Bodie Mining District of California - First published in 1986, it has been unavailable since those days and sheds important light on this famous mining area. Included are the history, characteristics and locations of numerous old mines around the ghost town of Bodie. 64 ppgs, 8.99

Geology and Mineral Resources of the Gasquet Quadrangle of California-Oregon - First published in 1953, it has been unavailable for over a century and sheds important light on the geological features and mineral resources of this portion of Northern California and Southern Oregon. 80 ppgs, 9.99

Alaska Mining Books

Ore Deposits of the Willow Creek Mining District, Alaska - Unavailable since 1954, this hard to find publication includes valuable insights into the Willow Creek Mining District near Hatcher Pass in Alaska. The publication includes insights into the history, geology and locations of the well known mines in the area, including the Gold Cord, Independence, Fern, Mabel, Lonesome, Snowbird, Schroff-O'Neil, High Grade, Marion Twin, Thorpe, Webfoot, Kelly-Willow, Lane, Holland and others. **8.5" X 11", 96 ppgs. Retail Price: $9.99**

The Juneau Gold Belt of Alaska - Unavailable since 1906, this hard to find publication includes valuable insights into the gold mines around Juneau, Alaska. The publication includes important details into the history, geology and locations of the well known gold mines and prospects in the area, including those around Windham Bay, Holkham Bay, Port Snettisham, on Grindstone and Rhine Creeks, Gold Creek, Douglas Island, Salmon Creek, Lemon Creek, Nugget Creek, from the Mendenhall River to Berners Bay, McGinnis Creek, Montana Creek, Peterson Creek, Windfall Creek, the Eagle River, Yankee Basin, Yankee Curve, Kowee Creek and elsewhere. Not only are gold placer mines included, but also hardrock gold mines. **8.5" X 11", 224 ppgs. Retail Price: $19.99**

Mining in the Jumbo Basin of Alaska - Unavailable since 1953, this hard to find publication includes valuable insights into the mines and geology of the Jumbo Basin. The publication includes important details into the history, geology and locations of the well known gold mines and prospects in the famous Jumbo Basin Mining Region of Alaska. 72 ppgs, 9.99

The Rampart Placer Gold Region of Alaska - Unavailable since 1906, this hard to find publication includes valuable insights into the placer gold mines of the Rampart Mining Region. The publication includes important details into the history, geology and locations of the well known gold mines and prospects in the famous Rampart Mining Region of Alaska. 78 ppgs, 10.99

Arizona Mining Books

Mines and Mining in Northern Yuma County Arizona - Originally published in 1911, this important publication on Arizona Mining has not been available for over a hundred years. Included are rare insights into the gold, silver, copper and quicksilver mines of Yuma County, Arizona together with hard to find maps and photographs. Some of the mines and mining districts featured include the Planet Copper Mine, Mineral Hill, the Clara Consolidated Mine, Viati Mine, Copper Basin prospect, Bowman Mine, Quartz King, Billy Mack, Carnation, the Wardwell and Osbourne, Valensuella Copper, the Mariquita, Colonial Mine, the French American, the New York-Plomosa, Guadalupe, Lead Camp, Mudersbach Copper Camp, Yellow Bird, the Arizona Northern (Salome Strike), Bonanza (Harqua Hala), Golden Eagle, Hercules, Socorro and others. **8.5" X 11", 144 ppgs. Retail Price: $11.99**

The Aravaipa and Stanley Mining Districts of Graham County Arizona - Originally published in 1925, this important publication on Arizona Mining has not been available for nearly ninety years. Included are rare insights into the gold and silver mines of these two important mining districts, together with hard to find maps. **8.5" X 11", 140 ppgs. Retail Price: $11.99**

Gold in the Gold Basin and Lost Basin Mining Districts of Mohave County, Arizona - This volume contains rare insights into the geology and gold mineralization of the Gold Basin and Lost Basin Mining Districts of Mohave County, Arizona that will be of benefit to miners and prospectors. Also included is a significant body of information on the gold mines and prospects of this portion of Arizona. This volume is lavishly illustrated with rare photos and mining maps. **8.5" X 11", 188 ppgs. Retail Price: $19.99**

Mines of the Jerome and Bradshaw Mountains of Arizona - This important publication on Arizona Mining has not been available for ninety years. This volume contains rare insights into the geology and ore deposits of the Jerome and Bradshaw Mountains of Arizona that will be of benefit to miners and prospectors who work those areas. Included is a significant body of information on the mines and prospects of the Verde, Black Hills, Cherry Creek, Prescott, Walker, Groom Creek, Hassayampa, Bigbug, Turkey Creek, Agua Fria, Black Canyon, Peck, Tiger, Pine Grove, Bradshaw, Tintop, Humbug and Castle Creek Mining Districts. This volume is lavishly illustrated with rare photos and mining maps. **8.5" X 11", 218 ppgs. Retail Price: $19.99**

The Ajo Mining District of Pima County Arizona - This important publication on Arizona Mining has not been available for nearly seventy years. This volume contains rare insights into the geology and mineralization of the Ajo Mining District in Pima County, Arizona and in particular the famous New Cornelia Mine. **8.5" X 11", 126 ppgs. Retail Price: $11.99**

Mining in the Santa Rita and Patagonia Mountains of Arizona - Originally published in 1915, this important publication on Arizona Mining has not been available for nearly a century. Included are rare insights into hundreds of gold, silver, copper and other mines in this famous Arizona mining area. Details include the locations, geology, history, production and other facts of the mines of this region. **8.5" X 11", 394 ppgs. Retail Price: $24.99**

Mining in the Bisbee Quadrangle of Arizona - Originally published in 1906, this important publication on Arizona Mining has not been available for nearly a century. Included are rare insights into hundreds of gold, silver, copper and other mines in this famous Arizona mining area. Details include the locations, geology, history, production and other facts of the mines of this important mining region. **8.5" X 11", 188 ppgs. Retail Price: $14.99**

Placer Gold Mining in Arizona - Unavailable since 1922, this hard to find publication includes valuable insights into the placer gold mines of the Arizona. Originally released as "Placer Gold of Arizona", despite its small size, this publication includes important details into the history, geology and locations of the well known placer gold mines and prospects in the State of Arizona. 48 ppgs, 8.99

Gold and Copper Mining near Payson, Arizona - Written in 1915, this hard to find publication includes valuable insights into the gold and copper mining industry of Arizona. Highlighted here are the gold and copper mines near Payson, Arizona. 68 ppgs, 8.99

Lode Gold Mining in Arizona - Unavailable since 1934, this hard to find publication, originally released as "Arizona Lode Gold Mines and Gold Mining" includes valuable insights into the gold mining industry of Arizona. Included are valuable insights into over 150 hardrock gold mines in over 30 different mining districts in Arizona. 278 ppgs, 21.99

Mining in the Dragoon Quadrangle of Cochise County, Arizona - Unavailable since 1964, this hard to find publication includes valuable insights into the mines of the Dragoon Quadrangle Mining Region. The publication includes important details into the history, geology and locations of the well known mines and prospects in this famous mining region of Arizona. 224 ppgs., 19.99

Directory of Operating Mines in Arizona in 1915 - Unavailable since 1916, this hard to find publication includes valuable insights into the mines of Arizona. This small publication includes a complete list of the mines that were operating in the State of Arizona during 1915 and includes details such as general location, owners and some basic facts about each mining operation. 52 ppgs. 8.99

Arizona Ore Deposits - Unavailable since 1938, this hard to find publication includes valuable insights into some ore deposits of Arizona. Included are valuable insights into the formation and characteristics of valuable ore deposits in the Jerome, Miami, Inspiration, Clifton, Morenci, Ray, Ajo, Eureka, Tombstone and Magma mining districts. Included are details into some of the major gold, silver and copper mines of these important Arizona mining areas. 160 ppgs, 14.99

A History of Butte Montana: The World's Greatest Mining Camp - First published in 1900 by H.C. Freeman, this important publication sheds a bright light on one of the most important mining areas in the history of The West. Together with his insights, as well as rare photographs of the periods, Harry Freeman describes Butte and its vicinity from its early beginnings, right up to its flush years when copper flowed from its mines like a river. At the time of publication, Butte, Montana was known worldwide as "The Richest Mining Spot On Earth" and produced not only vast amounts of copper, but also silver, gold and other metals from its mines. Freeman illustrates, with great detail, the most important mines in the vicinity of Butte, providing rare details on their owners, their history and most importantly, how the mines operated and how their treasures were extracted. Of particular interest are the dozens of rare photographs that depict mines such as the famous Anaconda, the Silver Bow, the Smoke House, Moose, Paulin, Buffalo, Little Minah, the Mountain Consolidated, West Greyrock, Cora, the Green Mountain, Diamond, Bell, Parnell, the Neversweat, Nipper, Original and many others.   **8.5" X 11", 142 ppgs. Retail Price: $12.99**

The Butte Mining District of Montana - This important publication on Montana Mining has not been available for over a century. Included are rare insights into the gold, copper and silver mines of Butte, Montana together with hard to find maps and photographs. Some of the topics include the early history of gold, silver and copper mining in the Butte area, insight into the geology of its mining areas, the local distribution of gold, silver and copper ores, as well their composition and how to identify them. Also included are detailed facts about the mines in the Butte Mining District, including the famous Anaconda Mine, Gagnon, Parrot, Blue Vein, Moscow, Poulin, Stella, Buffalo, Green Mountain, Wake Up Jim, the Diamond-Bell Group, Mountain Consolidated, East Greyrock, West Greyrock, Snowball, Corra, Speculator, Adirondack, Miners Union, the Jessie-Edith May Group, Otisco, Iduna, Colorado, Lizzie, Cambers, Anderson, Hesperus, Preferencia and dozens of others. **8.5" X 11", 298 ppgs. Retail Price: $24.99**

Mines of the Helena Mining Region of Montana - This important publication on Montana Mining has not been available for over a century. Included are rare insights into the gold, copper and silver mines of the vicinity of Helena, Montana, including the Marysville Mining District, Elliston Mining District, Rimini Mining District, Helena Mining District, Clancy Mining District, Wickes Mining District, Boulder and Basin Mining Districts and the Elkhorn Mining District. Some of the topics include the early history of gold, silver and copper mining in the Helena area, insight into the geology of its mining areas, the local distribution of gold, silver and copper ores, as well their composition and how to identify them. Also included are detailed facts, history, geology and locations of over one hundred gold, silver and copper mines in the area . **8.5" X 11", 162 ppgs, Retail Price: $14.99**

Mines and Geology of the Garnet Range of Montana - This important publication on Montana Mining has not been available for over a century. Included are rare insights into the gold, copper and silver mines of the vicinity of this important mining area of Montana. Some of the topics include the early history of gold, silver and copper mining in the Garnet Mountains, insight into the geology of its mining areas, the local distribution of gold, silver and copper ores, as well their composition and how to identify them. Also included are detailed facts, history, geology and locations of numerous gold, silver and copper mines in the area . **8.5" X 11", 100 ppgs, Retail Price: $11.99**

Mines and Geology of the Philipsburg Quadrangle of Montana - This important publication on Montana Mining has not been available for over a century. Included are rare insights into the gold, copper and silver mines of the vicinity of this important mining area of Montana. Some of the topics include the early history of gold, silver and copper mining in the Philipsburg Quadrangle, insight into the geology of its mining areas, the local distribution of gold, silver and copper ores, as well their composition and how to identify them. Also included are detailed facts, history, geology and locations of over one hundred gold, silver and copper mines in the area **8.5" X 11", 290 ppgs, Retail Price: $24.99**

Geology of the Marysville Mining District of Montana - Included are rare insights into the mining geology of the Marysville Mining District. Some of the topics include the early history of gold, silver and copper mining in the area, insight into the geology of its mining areas, the local distribution of gold, silver and copper ores, as well their composition and how to identify them. Also included are detailed facts, history, geology and locations of gold, silver and copper mines in the area **8.5" X 11", 198 ppgs, Retail Price: $19.99**

The Geology and Mines of Northern Idaho and North Western Montana - See listing under Idaho.

The History of Gold Dredging in Montana - Unavailable since 1916, this important publication was originally published by the Us Bureau of Mines and has been unavailable for a century. A century and more ago, giant dredging machines dug in Montana's rivers and creeks in search of illusive golden riches. First appearing in California in the 1850's, gold dredges finally reached their peak of development in Siberia and New Zealand before becoming popular again in the United States. This book offers a unique historical perspective on the gold dredges that once operated in Montana. This book on Montana mining history is lavishly illustrated with dozens of rare historic photos gold dredges that once operated in Montana, as well as hard to locate plans on how these dredges were designed. 120 ppgs., 11.99

Nevada Mining Books

The Bull Frog Mining District of Nevada - Unavailable since 1910, this publication was originally compiled by the United States Department of Interior. This volume also includes important insights into the geologic formations, faults and other aspects of economic geology in this Nevada mining district. Of particular interest are the fine details on many mines in the area, including their locations, histories, development and mineralization. Some of the mines featured include the National Bank Mine, Providence, Gibraltor, Tramps, Denver, Original Bullfrog, Gold Bar, Mayflower, Homestake-King and other mines and prospects. **8.5" X 11", 152 ppgs, Retail Price: $14.99**

History of the Comstock Lode - Unavailable since 1876, this publication was originally released by John Wiley & Sons. This volume also includes important insights into the famous Comstock Lode of Nevada that represented the first major silver discovery in the United States. During its spectacular run, the Comstock produced over 192 million ounces of silver and 8.2 million ounces of gold. Not only did the Comstock result in one of the largest mining rushes in history and yield immense fortunes for its owners, but it made important contributions to the development of the State of Nevada, as well as neighboring California. Included here are important details on not only the early development and history of the Comstock, but also rare early insight into its mines, ore and its geology. **8.5" X 11", 244 ppgs, Retail Price: $19.99**

The Pioche Mining District of Nevada - First published in 1932, it has been unavailable for over a century and sheds important light on the mining history of Nevada. Some of the topics include the history of mining in this district, as well as the characteristics of its mineral and ore deposits. Also included are insights into the history, production, characteristics and locations of numerous mines in the area. Some of the mines include the Combined Metals, Pioche, Ely Valley, No. 10, Poorman, Wide Awake, Alps, Prince, Virginia Louise, Half Moon, Abe Lincoln, Fairview, Bristol Silver, National, Vesuvius, Inman, Tempest, Hillside, Jackrabbit, Lucky Star, Fortuna, Mendha, Manhattan, Hamburg, Comet, Lyndon and others. 108 ppgs 10.99

The Yerington Mining District of Nevada - First published in 1932, it has been unavailable for over a century and sheds important light on the mining history of Nevada. Some of the topics include the history of mining in this district, as well as the characteristics of its mineral and ore deposits. Also included are insights into the history, production, characteristics and locations of numerous mines in the area. Some of the mines include the Bluestone, Mason Valley, Malachite, McConnell, Greenwood, Western Nevada, Ludwig, Douglas Hill, Casting Copper, Montana-Yerington, Empire, Jim Beatty, Terry and McFarland, Blue Jay and others. 92 ppgs, 10.99

The Genesis of the Ores of Tonopah Nevada - Unavailable since 1918, this hard to find publication includes valuable insights into the gold mines around Tonopah, Nevada. The publication includes important details into the geology of mines in the Tonopah Mining District of Nevada. 90 ppgs, 10.99

Mining Camps of Elko, Lander and Eureka Counties Nevada - Unavailable since 1910, this hard to find publication includes valuable insights into the mining camps of Elko, Lander and Eureka Counties, Nevada. The publication includes important details into the history of mines and mining in these three Nevada counties. 154 ppgs, 12.99

Ore Deposits of the Bullfrog Quadrangle - Unavailable since 1964 and released as "Geology of Bullfrog Quadrangle and Ore Deposits Related to Bullfrog Hills Caldera, Nye County, Nevada and Inyo County, California". The publication includes important details into the geology of mines in the Bullfrog Quadrangle of Nye County, Nevada and Inyo County, California. 52 ppgs, 9.99

Mining in Eureka County Nevada - Unavailable since 1879, this hard to find publication includes valuable insights into the early mining history off Eureka County, Nevada. The publication includes important details into the early history of the mines of Eureka County, as well as their development, production and how their ores were treated. Also included are details on the 1872 Mining Act, as well as the local rules, regulations and customs of the miners in Eureka County. 134 ppgs, 12.99

Colorado Mining Books

Ores of The Leadville Mining District - Unavailable since 1926, this publication was originally compiled by the United States Department of Interior. This volume also includes important insights into the ores and mineralization of the Leadville Mining District in Colorado. Topics include historic ore prospecting methods, local geology, insights into ore veins and stockworks, the local trend and distribution of ore channels, reverse faults, shattered rock above replacement ore bodies, mineral enrichment in oxidized and sulphide zones and more. **8.5" X 11", 66 ppgs, Retail Price: $8.99**

Mining in Colorado - Unavailable since 1926, this publication was originally compiled by the United States Department of Interior. This volume also includes important insights into the mining history of Colorado from its early beginnings in the 1850's right up to the mid 1920's. Not only is Colorado's gold mining heritage included, but also its silver, copper, lead and zinc mining industry. Each mining area is treated separately, detailing the development of Colorado's mines on a county by county basis. **8.5" X 11", 284 ppgs, Retail Price: $19.99**

Gold Mining in Gilpin County Colorado - Unavailable since 1876, this publication was originally compiled by the Register Steam Printing House of Central City, Colorado. A rare glimpse at the gold mining history and early mines of Gilpin County, Colorado from their first discovery in the 1850's up to the "flush years" of the mid 1870's. Of particular interest is the history of the discovery of gold in Gilpin County and details about the men who made those first strikes. Special focus is given to the early gold mines and first mining districts of the area, many of which are not detailed in other books on Colorado's gold mining history. **8.5" X 11", 156 ppgs, Retail Price: $12.99**

Mining in the Gold Brick Mining District of Colorado - Important insights into the history of the Gold Brick Mining District, as well as its local geography and economic geology. Also included are the histories and locations of historic mines in this important Colorado Mining District, including the Cortland, Carter, Raymond, Gold Links, Sacramento, Bassick, Sandy Hook, Chronicle, Grand Prize, Chloride, Granite Mountain, Lucille, Gray Mountain, Hilltop, Maggie Mitchell, Silver Islet, Revenue, Roosevelt, Carbonate King and others. In addition to hardrock mining, are also included are details on gold placer mining in this portion of Colorado. **8.5" X 11", 140 ppgs, Retail Price: $12.99**

Ore Deposits of the London Fault of Colorado - First published in 1941, it has been unavailable since those days and sheds important light on the mines and mineral deposits of the London Fault in Central Colorado's Alma Mining District. This publication sheds important light on the gold veins and lead-silver deposits of the Alma Mining District. Included are geologic details on the London Mine, American Mine, Havigorst Tunnel, Ophir Mine, Mosher Tunnel, London-Butte Mine, Venture Shaft, Hard-To-Beat Mine, Oliver Twist Tunnel, Sacramento Mine, Mudsill Mine, Sherwood Mine, Wagner, Barcoe Tunnel and other mines in this important mining region. 110 ppgs., 10.99

The Mines of Colorado - First published in 1867, it has been unavailable since those days and sheds important light on Colorado's early mining history. Written shortly after the events took place, this publication sheds important light on the Pike's Peak Gold Rush, the discovery of gold on Ralston Creek and Dry Creek in the 1850's, as well as details on the first wave of miners into Colorado and their trials and tribulations as they crossed the Great Plains. Also included are details on early discoveries of lode gold in the mountainous regions of Colorado, details on the early mines hardrock and placer mines, and much more. It is a veritable treasure trove on Colorado's early mining history and will be of great importance to anyone who is interested in the mining of gold or other minerals in Colorado, as well as those interested in the history of the state. 478 ppgs., 29.99

The La Plata Mining District of Colorado - Originally titled "Geology and Ore Deposits in the Vicinity of the La Plata District of Colorado" and first published in 1949, it has been unavailable since those days and sheds important light on the mines and mineral deposits of the La Plata Mining District of Colorado. 214 ppgs., 19.99

Washington Mining Books

The Republic Mining District of Washington - Unavailable since 1910, this important publication was originally published by the Washington Geologic Survey and has been unavailable for a century. Topics include the geology, rock formations and the formation of ore deposits in this important mining area of Washington State. Also included are hard to find details on the geology, history and locations of dozens of mines in the area. Some of the mines featured include the New Republic Mine, Ben Hur, Morning Glory, the South Republic Mine, Quilp, Surprise, Black Tail, Lone Pine, San Poil, Mountain Lion, Tom Thumb, Elcaliph and many others. **8.5" X 11", 94 ppgs, Retail Price: $10.99**

The Myers Creek and Nighthawk Mining Districts of Washington - Unavailable since 1911, this important publication was originally published by the Washington Geologic Survey and has been unavailable for a century. Topics include the geology, rock formations and the formation of ore deposits in these important mining areas of Washington State. Also included are hard to find details on the geology, history and locations of dozens of mines in the area. Some of the mines featured include the Grant Mine, Monterey, Nip and Tuck, Myers Creek, Number Nine, Neutral, Rainbow, Aztec, Crystal Butte, Apex, Butcher Boy, Molson, Mad River, Olentangy, Delate, Kelsey, Golden Chariot, Okanogan, Ohio, Forty-Ninth Parallel, Nighthawk, Favorite, Little Chopaka, Summit, Number One, California, Peerless, Caaba, Prize Group, Ruby, Mountain Sheep, Golden Zone, Rich Bar, Similkameen, Kimberly, Triune, Hiawatha, Trinity, Hornsilver, Maquae, Bellevue, Bullfrog, Palmer Lake, Ivanhoe, Copper World and many others. **8.5" X 11", 136 ppgs, Retail Price: $12.99**

The Blewett Mining District of Washington - Unavailable since 1911, this important publication was originally published by the Washington Geologic Survey and has been unavailable for a century. Topics include the geology, rock formations and the formation of ore deposits in this important mining area of Washington State. Also included are hard to find details on the geology, history and locations of dozens of mines in the area. Some of the mines featured include the Washington Meteor, Alta Vista, Pole Pick, Blinn, North Star, Golden Eagle, Tip Top, Wilder, Golden Guinea, Lucky Queen, Blue Bell, Prospect, Homestake, Lone Rock, Johnson, and others. **8.5" X 11", 134 ppgs, Retail Price: $12.99**

Silver Mining In Washington - Unavailable since 1955, this important publication was originally published by the Washington Geologic Survey. Featured are the hard to find locations and details pertaining to Washington's silver mines. **8.5" X 11", 180 ppgs, Retail Price: $15.99**

The Mines of Snohomish County Washington - Unavailable since 1942, this important publication was originally published by the Washington Geologic Survey and has been unavailable for seventy years. Featured are details on a large number of gold, silver, copper, lead and other metallic mineral mines. Included are the locations of each historic mine, along with information on the commodity produced. **8.5" X 11", 98 ppgs, Retail Price: $10.99**

The Mines of Chelan County Washington - Unavailable since 1943, this important publication was originally published by the Washington Geologic Survey and has been unavailable for seventy years. Featured are details on a large number of gold, silver, copper, lead and other metallic mineral mines. Included are the locations of each historic mine, along with information on the commodity. **8.5" X 11", 88 ppgs, Retail Price: $9.99**

Metal Mines of Washington - Unavailable since 1921, this important publication was originally published by the Washington Geologic Survey and has been unavailable for nearly ninety years. Widely considered a masterpiece on the Washington Mining Industry, "Metal Mines of Washington" sheds light on the important details of Washington's early mining years. Featured are details on hundreds of gold, silver, copper, lead and other metallic mineral mines. Included are hard to find details on the mineral resources of this state, as well as the locations of historic mines. Lavishly illustrated with maps and historic photos and complete with a glossary to explain any technical terms found in the text, this is one of the most important works on mining in the State of Washington. No prospector or miner should be without it if they are interested in mining in Washington. **8.5" X 11", 396 ppgs, Retail Price: $24.99**

Gem Stones In Washington - Unavailable since 1949, this important publication was originally published by the Washington Geologic Survey and has been unavailable since first published. Included are details on where to find naturally occurring gem stones in the State of Washington, including quartz crystal, amethyst, smoky quartz, milky quartz, agates, bloodstone, carnelian, chert, flint, jasper, onyx, petrified wood, opal, fire opal, hyalite and others. **8.5" X 11", 54 ppgs, Retail Price: $8.99**

The Covada Mining District of Washington - Unavailable since 1913, this important publication was originally published by the Washington Geologic Survey and has been unavailable for a century. Topics include the geology, rock formations and the formation of ore deposits in this important mining area of Washington State. Also included are hard to find details on the geology, history and locations of dozens of mines in the area. Some of the mines featured include the Admiral, Advance, Algonkian, Big Bug, Big Chief, Big Joker, Black Hawk, Black Tail, Black Thorn, Captain, Cherokee Strip, Colorado, Dan Patch, Dead Shot, Etta, Good Ore, Greasy Run, Great Scott, Idora, IXL, Jay Bird, Kentucky Bell, King Solomon, Laurel, Laura S, Little Jay, Meteor, Neglected, Northern Light, Old Nell, Plymouth Rock, Polaris, Quandary, Reserve, Shoo Fly, Silver Plume, Three Pines, Vernie, White Rose and dozens of others. **8.5" X 11", 114 ppgs, Retail Price: $10.99**

The Index Mining District of Washington - Unavailable since 1912, this important publication was originally published by the Washington Geologic Survey and has been unavailable for a century. Topics include the geology, rock formations and the formation of ore deposits in this important mining area of Washington State. Also included are hard to find details on the geology, history and locations of dozens of mines in the area. Some of the mines featured include the Sunset, Non-Pareil, Ethel Consolidated, Kittaning, Merchant, Homestead, Co-operative, Lost Creek, Uncle Sam, Calumet, Florence-Rae, Bitter Creek, Index Peacock, Gunn Peak, Helena, North Star, Buckeye. Copper Bell, Red Cross and others. **8.5" X 11", 114 ppgs, Retail Price: $11.99**

Mining & Mineral Resources of Stevens County Washington - Unavailable since 1920, this important publication was originally published by the Washington Geologic Survey and has been unavailable for a century. Topics include the geology, rock formations and the formation of ore deposits in these important mining areas of Washington State. Also included are hard to find details on the geology, history and locations of hundreds of mines in the area. **8.5" X 11", 372 ppgs, Retail Price: $24.99**

The Mines and Geology of the Loomis Quadrangle Okanogan County, Washington - Unavailable since 1972, this important publication was originally published by the Washington Geologic Survey and has been unavailable for a century. Topics include the geology, rock formations and the formation of ore deposits in this important mining area of Washington State. Also included are hard to find details on the geology, history and locations of dozens of gold, copper, silver and other mines in the area. **8.5" X 11", 150 ppgs, Retail Price: $12.99**

The Conconully Mining District of Okanogan County Washington - Unavailable since 1973, this important publication was originally published by the Washington Geologic Survey and has been unavailable for a century. Topics include the geology, rock formations and the formation of ore deposits in this important mining area of Washington State, which also includes Salmon Creek, Blue Lake and Galena. Also included are hard to find details on the geology, mining history and locations of dozens of mines in the area. Some of the mines include Arlington, Fourth of July, Sonny Boy, First Thought, Last Chance, War Eagle-Peacock, Wheeler, Mohawk, Lone Star, Woo Loo Moo Loo, Keystone, Hughes, Plant-Callahan, Johnny Boy, Leuena, Gubser, John Arthur, Tough Nut, Homestake, Key and many others **8.5" X 11", 68 ppgs, Retail Price: $8.99**

Wyoming Mining Books

Mining in the Laramie Basin of Wyoming - Unavailable since 1909, this publication was originally compiled by the United States Department of Interior. Also included are insights into the mineralization and other characteristics of this important mining region, especially in regards to coal, limestone, gypsum, bentonite clay, cement, sand, clay and copper. **8.5" X 11", 104 ppgs, Retail Price: $11.99**

New Mexico Mining Books

The Mogollon Mining District of New Mexico - Unavailable since 1927, this important publication was originally published by the US Department of Interior and has been unavailable for 80 years. Topics include the geology, rock formations and the formation of ore deposits in this important mining area in New Mexico. Of particular focus is information on the history and production of the ore deposits in this area, their form and structure, vein filling, their paragenesis, origins and ore shoots, as well as oxidation and supergene enrichment. Also included are hard to find details, including the descriptions and locations of numerous gold, silver and other types of mines, including the Eureka, Pacific, South Alpine, Great Western, Enterprise, Buffalo, Mountain View, Floride, Gold Dust, Last Chance, Deadwood, Confidence, Maud S., Deep Down, Little Fanney, Trilby, Johnson, Alberta, Comet, Golden Eagle, Cooney, Queen, the Iron Crown, Eberle, Clifton, Andrew Jackson mine, Mascot and others. **8.5" X 11", 144 ppgs, Retail Price: $12.99**

The Percha Mining District of Kingston New Mexico - Unavailable since 1883, this important publication was originally published by the Kingston Tribune and has been unavailable for over one hundred and thirty five years. Having been written during the earliest years of gold and silver mining in the Percha Mining District, unlike other books on the subject, this work offers the unique perspective of having actually been written while the early mining history of this area was still being made. In fact, the work was written so early in the development of this area that many of the notable mines in the Percha District were less than a few years old and were still being operated by their original discoverers with the same enthusiasm as when they were first located. Included are hard to find details on the very earliest gold and silver mines of this important mining district near Kingston in Sierra County, New Mexico. **8.5" X 11", 68 ppgs, Retail Price: $9.99**

East Coast Mining Books

<u>The Gold Fields of the Southern Appalachians</u> - Unavailable since 1895, this important publication was originally published by the US Department of Interior and has been unavailable for nearly 120 years. Topics include the geology, rock formations and the formation of ore deposits in this important mining area of the American South. Of particular focus is information on the history and statistics of the ore deposits in this area, their form and structure and veins. Also included are details on the placer gold deposits of the region. The gold fields of the Georgian Belt, Carolinian Belt and the South Mountain Mining District of North Carolina are all treated in descriptive detail. Included are hard to find details, including the descriptions and locations of numerous gold mines in Georgia, North Carolina and elsewhere in the American South. Also included are details on the gold belts of the British Maritime Provinces and the Green Mountains. **8.5" X 11", 104 ppgs, Retail Price: $9.99**

Gold Rush Tales Series

Millions in Siskiyou County Gold - In this first volume of the "Gold Rush Tales" series, leading mining historian and editor Kerby Jackson, introduces us to the story of how millions of dollars worth of gold was discovered in Siskiyou County during the California Gold Rush. Lavishly illustrated with photos from the 19th Century, this hard to find information was first published in 1897 and sheds important light onto the gold rush era in Siskiyou County, California and the experiences of the men who dug for the gold and actually found it. **8.5" X 11", 82 ppgs, Retail Price: $9.99**

The California Rand in the Days of '49 - In this second volume of the "Gold Rush Tales" series, leading mining historian and editor Kerby Jackson, introduces us to four tales from the California Gold Rush. Lavishly illustrated with photos from the 19th Century, this hard to find information was first published in 1890's and includes the stories of "California's Rand", details about Chinese miners, how one early miner named Baker struck it rich and also the story of Alphonzo Bowers, who invented the first hydraulic gold dredge. **8.5" X 11", 54 ppgs, Retail Price: $9.99**

More Mining Books

Prospecting and Developing A Small Mine - Topics covered include the classification of varying ores, how to take a proper ore sample, the proper reduction of ore samples, alluvial sampling, how to understand geology as it is applied to prospecting and mining, prospecting procedures, methods of ore treatment, the application of drilling and blasting in a small mine and other topics that the small scale miner will find of benefit. **8.5" X 11", 112 ppgs, Retail Price: $11.99**

Timbering For Small Underground Mines - Topics covered include the selection of caps and posts, the treatment of mine timbers, how to install mine timbers, repairing damaged timbers, use of drift supports, headboards, squeeze sets, ore chute construction, mine cribbing, square set timbering methods, the use of steel and concrete sets and other topics that the small underground miner will find of benefit. This volume also includes twenty eight illustrations depicting the proper construction of mine timbering and support systems that greatly enhance the practical usability of the information contained in this small book. **8.5" X 11", 88 ppgs. Retail Price: $10.99**

Timbering and Mining - A classic mining publication on Hard Rock Mining by W.H. Storms. Unavailable since 1909, this rare publication provides an in depth look at American methods of underground mine timbering and mining methods. Topics include the selection and preservation of mine timbers, drifting and drift sets, driving in running ground, structural steel in mine workings, timbering drifts in gravel mines, timbering methods for driving shafts, positioning drill holes in shafts, timbering stations at shafts, drainage, mining large ore bodies by means of open cuts or by the "Glory Hole" system, stoping out ore in flat or low lying veins, use of the "Caving System", stoping in swelling ground, how to stope out large ore bodies, Square Set timbering on the Comstock and its modifications by California miners, the construction of ore chutes, stoping ore bodies by use of the "Block System", how to work dangerous ground, information on the "Delprat System" of stoping without mine timbers, construction and use of headframes and much more. This volume provides a reference into not only practical methods of mining and timbering that may be employed in narrow vein mining by small miners today, but also rare insights into how mines were being worked at the turn of the 19th Century. **8.5" X 11", 288 ppgs. Retail Price: $24.99**

A Study of Ore Deposits For The Practical Miner - Mining historian Kerby Jackson introduces us to a classic mining publication on ore deposits by J.P. Wallace. First published in 1908, it has been unavailable for over a century. Included are important insights into the properties of minerals and their identification, on the occurrence and origin of gold, on gold alloys, insights into gold bearing sulfides such as pyrites and arsenopyrites, on gold bearing vanadium, gold and silver tellurides, lead and mercury tellurides, on silver ores, platinum and iridium, mercury ores, copper ores, lead ores, zinc ores, iron ores, chromium ores, manganese ores, nickel ores, tin ores, tungsten ores and others. Also included are facts regarding rock forming minerals, their composition and occurrences, on igneous, sedimentary, metamorphic and intrusive rocks, as well as how they are geologically disturbed by dikes, flows and faults, as well as the effects of these geologic actions and why they are important to the miner. Written specifically with the common miner and prospector in mind, the book will help to unlock the earth's hidden wealth for you and is written in a simple and concise language that anyone can understand. **8.5" X 11", 366 ppgs. Retail Price: $24.99**

Mine Drainage - Unavailable since 1896, this rare publication provides an in depth look at American methods of underground mine drainage and mining pump systems. This volume provides a reference into not only practical methods of mining drainage that may be employed in narrow vein mining by small miners today, but also rare insights into how mines were being worked at the turn of the 19th Century. **8.5" X 11", 218 ppgs. Retail Price: $24.99**

Fire Assaying Gold, Silver and Lead Ores - Unavailable since 1907, this important publication was originally published by the Mining and Scientific Press and was designed to introduce miners and prospectors of gold, silver and lead to the art of fire assaying. Topics include the fire assaying of ores and products containing gold, silver and lead; the sampling and preparation of ore for an assay; care of the assay office, assay furnaces; crucibles and scorifiers; assay balances; metallic ores; scorification assays; cupelling; parting' crucible assays, the roasting of ores and more. This classic provides a time honored method of assaying put forward in a clear, concise and easy to understand language that will make it a benefit to even beginners. **8.5" X 11", 96 ppgs. Retail Price: $11.99**

Methods of Mine Timbering - Originally published in 1896, this important publication on mining engineering has not been available for nearly a century. Included are rare insights into historical methods of timbering structural support that were used in underground metal mines during the California that still have a practical application for the small scale hardrock miner of today. **8.5" X 11", 94 ppgs. Retail Price: $10.99**

The Enrichment of Copper Sulfide Ores - First published in 1913, it has been unavailable for over a century. Topics include the definition and types of ore enrichment, the oxidation of copper ores, the precipitation of metallic sulfides. Also included are the results of dozens of lab experiments pertaining to the enrichment of sulfide ores that will be of interest to the practical hard rock mine operator in his efforts to release the metallic bounty from his mine's ore. **8.5" X 11", 92 ppgs. Retail Price: $9.99**

A Study of Magmatic Sulfide Ores - Unavailable since 1914, this rare publication provides an in depth look at magmatic sulfide ores. Some of the topics included are the definition and classification of magmatic ores, descriptions of some magmatic sulfide ore deposits known at the time of publication including copper and nickel bearing pyrrohitic ore bodies, chalcopyrite-bornite deposits, pyritic deposits, magnetite-ileminite deposits, chromite deposits and magmatic iron ore deposits. Also included are details on how to recognize these types of ore deposits while prospecting for valuable hardrock minerals. **8.5" X 11", 138 ppgs. Retail Price: $11.99**

The Cyanide Process of Gold Recovery - Unavailable since 1894 and released under the name "The Cyanide Process: Its Practical Application and Economical Results", this rare publication provides an in depth look at the early use of cyanide leaching for gold recovery from hardrock mine ores. This volume provides a reference into the early development and use of cyanide leaching to recover gold. **8.5" X 11", 162 ppgs. Retail Price: $14.99**

California Gold Milling Practices - Unavailable since 1895 and released under the name "California Gold Practices", this rare publication provides an in depth look at early methods of milling used to reduce gold ores in California during the late 19th century. This volume provides a reference into the early development and use of milling equipment during the earliest years of the California Gold Rush up to the age of the Industrial Revolution. Much of the information still applies today and will be of use to small scale miners engaging in hardrock mining. **8.5" X 11", 104 ppgs. Retail Price: $10.99**

Leaching Gold and Silver Ores With The Plattner and Kiss Processes - Mining historian Kerby Jackson introduces us to a classic mining publication on the evaluation and examination of mines and prospects by C.H. Aaron. First published in 1881, it has been unavailable for over a century and sheds important light on the leaching of gold and silver ores with the Plattner and Kiss processes. **8.5" X 11", 204 ppgs. Retail Price: $15.99**

The Metallurgy of Lead and the Desilverization of Base Bullion - First published in 1896, it has been unavailable for over a century and sheds important light on the the recovery of silver from lead based ores. Some of the topics include the properties of lead and some of its compounds, lead ores such as galenite, anglesite, cerussite and others, the distribution of lead ores throughout the United States and the sampling and assaying of lead ores. Also covered is the metallurgical treatment of lead ores, as well as the desilverization of lead by the Pattinson Process and the Parkes Process. Hofman's text has long been considered one of the most important early works on the recovery of silver from lead based ores. 8.5" X 11", 452 ppgs. **Retail Price: $29.99**

Ore Sampling For Small Scale Miners - First published in 1916, it has been unavailable for over a century and sheds important light on historic methods of ore sampling in hardrock mines. Topics include how to take correct ore samples and the conditions that affect sampling, such as their subdivision and uniformity. Particular detail is given to methods of hand sampling ore bodies by grab sample, pipe sample and coning, as well as sampling by mechanical methods. Also given are insights into the screening, drying and grinding processes to achieve the most consistent sample results and much more. 8.5" X 11", 124 ppgs. **Retail Price: $12.99**

The Extraction of Silver, Copper and Tin from Ores - First published in 1896, it has been unavailable for over a century and sheds important light on how historic miners recovered silver, copper and tin from their mining operations. The book is split into three sections, including a discussion on the Lixiviation of Silver Ores, the mining and treatment of copper ores as practiced at Tharsis, Spain and the smelting of tin as it was practiced by metallurgists at Pulo Brani, Singapore. Also included is an overview and analysis of these historic metal recovery methods that will be of benefit to those interested in the extraction of silver, copper and tin from small mines. 8.5" X 11", 118 ppgs. **Retail Price: $14.99**

The Roasting of Gold and Silver Ores - First published in 1880, it has been unavailable for over a century and sheds important light on how historic miners recovered gold and silver rom their mining operations. Topics include details on the most important silver and free milling gold ores, methods of desulphurization of ores, methods of deoxidation, the chlorination of ores, methods and details on roasting gold and silver ores, notes on furnaces and much more. Also included are details on numerous methods of gold and silver recovery, including the Ottokar Hofman's Process, the Patera Process, Kiss Process, Augustin Process, Ziervogel Process and others. 8.5" X 11", 178 ppgs. **Retail Price: $19.99**

The Examination of Mines and Prospects - First published in 1912, it has been unavailable for over a century and sheds important light on how to examine and evaluate hardrock mines, prospects and lode mining claims. Sections include Mining Examinations, Structural Geology, Structural Features of Ore Deposits, Primary Ores and their Distribution, Types of Primary Ore Deposits, Primary Ore Shoots, The Primary Alteration of Wall Rocks, Alterations by Surface Agencies, Residual Ores and their Distribution, Secondary Ores and Ore Shoots and Vein Outcrops. This hard to find information is a must for those who are interested in owning a mine or who already own a lode mining claim and wish to succeed at quartz mining. 8.5" X 11", 250 ppgs. **Retail Price: $19.99**

Garnets: Their Mining, Milling and Utilization - First published in 1925, it has been unavailable since those days and sheds important light on the mining, milling and utilization of garnets. Included are details on the characteristics of garnets, where they are found and how they were mined. 78 ppgs, 10.99

Gemstones and Precious Stones of North America - Leading mining historian Kerby Jackson introduces us to a classic mining publication on the gems and precious stones of the United States, Canada and mexico. First published in 1890, it has been unavailable since those days and sheds important light on the gems and precious stones that may be found in North America. Included are chapters on diamonds, corundum, sapphire, ruby, topaz, emerald, disapore, spinel, turquoise, tourmaline, garnets, beyrl, peridot, zircon, quartz crystals, feldspars, pearls and many others. Included are details on where these gems and precious stones may be found throughout North America, as well as their characteristics. 360 ppgs, 24.99

Mining Camps and Mining Districts - First released in 1885 by Charles Howard Shinn under the title "Mining Camps: A Study in American Frontier Government", this publication offers a unique look at how early gold miners established their own forms of representative government during the California Gold Rush. Drawing on the the early mining codes of mideviel German miners in the Harz Mountains, on the mining customs of the Cornish tin miners and early Spanish mining laws introduced into California, the miners established the first governments in the American West. 340 ppgs, 24.99

BLM Field Handbook for Mineral Examiners - Leading mining historian Kerby Jackson introduces us to a classic mining publication on mine evaluation. First published in 1962, this work sheds important light on the techniques of BLM Mineral Examiners to perform validity on mining claims. 132 ppgs, 10.99

Six Months In The Gold Mines During The California Gold Rush - Unavailable since 1850, this important work is a first hand account of one "49'ers" personal experience during the great California Gold Rush, shedding important light on one of the most exciting periods in the history of not only California, but also the world. Compiled from journals written between 1847 and 1849 by E. Gould Buffum, a native of New York, "Six Months In The Gold Mines During The California Gold Rush" offers a rare look into the day to day lives of the people who came to California to work in her gold mines when the state was still a great frontier. **8.5" X 11", 290 ppgs. Retail Price: $19.99**

The Discovery of Gold in Australia - **First published in 1852, it has been unavailable since those days and sheds important light on Australia's gold mining history. Included are rare communications between British agents and the British Crown when gold was first discovered in Australia in 1851. This rare text contains hard to find details on Australia's first mining camps and Britain's early attempts to provide for the orderly regulation of gold mines in that part of the world. Also of interest are hard to find extracts of articles that appeared in the early colonial newspapers that did their best to report on Australia's gold rush as it took place.**
102 ppgs, 10.99

www.ingramcontent.com/pod-product-compliance
Lightning Source LLC
Chambersburg PA
CBHW080712190526
45169CB00006B/2339